Caution! Wireless Networking: Preventing a Data Disaster

Caution! Wireless Networking: Preventing a Data Disaster

Jack McCullough

Wiley Publishing, Inc.

Caution! Wireless Networking: Preventing a Data Disaster

Published by
Wiley Publishing, Inc.
111 River Street
Hoboken, N.J. 07030
www.wiley.com

Copyright © 2004 by Wiley Publishing, Inc., Indianapolis, Indiana

Published by Wiley Publishing, Inc., Indianapolis, Indiana

Published simultaneously in Canada

Library of Congress Cataloging-in-Publication Data

ISBN: 0-7645-7213-X

Manufactured in the United States of America

10 9 8 7 6 5 4 3 2

1MA/SS/QZ/QU/IN

No part of this publication may be reproduced, stored in a retrieval system or transmitted in any form or by any means, electronic, mechanical, photocopying, recording, scanning or otherwise, except as permitted under Sections 107 or 108 of the 1976 United States Copyright Act, without either the prior written permission of the Publisher, or authorization through payment of the appropriate per-copy fee to the Copyright Clearance Center, 222 Rosewood Drive, Danvers, MA 01923, (978) 750-8400, fax (978) 750-4744. Requests to the Publisher for permission should be addressed to the Legal Department, Wiley Publishing, Inc., 10475 Crosspoint Blvd., Indianapolis, IN 46256, (317) 572-3447, fax (317) 572-4355, E-Mail: brandreview@wiley.com.

LIMIT OF LIABILITY/DISCLAIMER OF WARRANTY: THE PUBLISHER AND THE AUTHOR MAKE NO REPRESENTATIONS OR WARRANTIES WITH RESPECT TO THE ACCURACY OR COMPLETENESS OF THE CONTENTS OF THIS WORK AND SPECIFICALLY DISCLAIM ALL WARRANTIES, INCLUDING WITHOUT LIMITATION WARRANTIES OF FITNESS FOR A PARTICULAR PURPOSE. NO WARRANTY MAY BE CREATED OR EXTENDED BY SALES OR PROMOTIONAL MATERIALS. THE ADVICE AND STRATEGIES CONTAINED HEREIN MAY NOT BE SUITABLE FOR EVERY SITUATION. THIS WORK IS SOLD WITH THE UNDERSTANDING THAT THE PUBLISHER IS NOT ENGAGED IN RENDERING LEGAL, ACCOUNTING, OR OTHER PROFESSIONAL SERVICES. IF PROFESSIONAL ASSISTANCE IS REQUIRED, THE SERVICES OF A COMPETENT PROFESSIONAL PERSON SHOULD BE SOUGHT. NEITHER THE PUBLISHER NOR THE AUTHOR SHALL BE LIABLE FOR DAMAGES ARISING HEREFROM. THE FACT THAT AN ORGANIZATION OR WEB SITE IS REFERRED TO IN THIS WORK AS A CITATION AND/OR A POTENTIAL SOURCE OF FURTHER INFORMATION DOES NOT MEAN THAT THE AUTHOR OR THE PUBLISHER ENDORSES THE INFORMATION THE ORGANIZATION OF WEB SITE MAY PROVIDE OR RECOMMENDATIONS IT MAY MAKE. FURTHER, READERS SHOULD BE AWARE THAT INTERNET WEB SITES LISTED IN THIS WORK MAY HAVE CHANGED OR DISAPPEARED BETWEEN WHEN THIS WORK WAS WRITTEN AND WHEN IT IS READ.

For general information on our other products and services or to obtain technical support, please contact our Customer Care Department within the U.S. at (800) 762-2974, outside the U.S. at (317) 572-3993 or fax (317) 572-4002.

Wiley also publishes its books in a variety of electronic formats. Some content that appears in print may not be available in electronic books.

Trademarks: Wiley and the Wiley Publishing logo are trademarks or registered trademarks of John Wiley and Sons, Inc. and/or its affiliates. All other trademarks are the property of their respective owners. Wiley Publishing, Inc. is not associated with any product or vendor mentioned in this book.

WILEY is a trademark of Wiley Publishing, Inc.

About the Author

Jack McCullough is the managing director of Razorwire Information Security Consulting. His technical expertise includes wireless and wired networks, computer security, physical security, programming, cryptography, and technical curriculum development. Jack's background includes ten years of experience in the IT field. He has held positions as IT director, operations manager, network administrator, programmer, and software trainer. A respected IT and security authority, he is frequently sought out for informational interviews by both broadcast and print media services.

Jack has authored books, magazine articles, and white papers on computer security; his written works have been translated into several languages. Many universities have used his books and white papers in information security courses, as have the governments of Australia, the Peoples Republic of China, Japan, Brazil, and Taiwan. Jack continues to actively research information security, discover new ways to exploit the weaknesses in networked systems, and determine best practices that enable the average computer user to address these threats in an efficient manner.

When he isn't writing about or researching technology, Jack teaches karate and self-defense under the watchful eyes of Sensei Floyd Burk, and Sensei Martha Burk at the Alpine Karate Academy in Alpine, California, and practices writing about himself in third-person.

Credits

Acquisitions Editor
Mike Roney

Project Editor
Cricket Krengel

Technical Editor
Greg Guntle

Copy Editor
Kim Heusel

Editorial Manager
Robyn Siesky

VP & Publisher
Barry Pruett

Vice President & Group Executive Publisher
Richard Swadley

Project Coordinator
Maridee Ennis

Graphics and Production Specialists
Karl Brandt
Jonelle Burns
Amanda Carter
Carrie A. Foster
Lauren Goddard
Jennifer Heleine

Quality Control Technicians
Susan Moritz
Carl William Pierce
Brian H. Walls

Proofreading
Christine Sabooni

Indexing
Johnna VanHoose

*To Cathy, for her support during this
and all of my other projects.*

Acknowledgments

It takes a lot of exceptional people to complete a project like this. First and foremost, a big thank you goes to my literary agent, David Fugate of Waterside productions, arguably the best agent working for technical authors today. I want to thank Mike Roney for sharing his vision for the book and hiring me to write it, and my development editor, Cricket Krengel, who expertly guided me through preparing this manuscript and put so much work into editing it — especially after the vision of the title changed and we had to rethink several chapters. Finally, thanks to the layout and production team at Wiley for putting this book together and for interpreting my scribbles into actual art that the reader can understand.

Contents at a Glance

Acknowledgments ix

Part I: Understanding the Threat 1
Chapter 1: A Brief Overview of Wi-Fi 3
Chapter 2: Network Fundamentals and Security Concerns 23
Chapter 3: The People behind the Problem 45
Chapter 4: Hijacking Wi-Fi 59
Chapter 5: More Wireless Attacks 77
Chapter 6: Wardriving 89
Chapter 7: Viruses and Wi-Fi 101

Part II: Protecting Yourself 123
Chapter 8: Technical Pitfalls and Solutions 125
Chapter 9: Wireless Privacy Concerns 147
Chapter 10: Encryption and Wi-Fi 167
Chapter 11: Securing Your WLAN 187
Chapter 12: Protecting Your Wi-Fi Data 217

Appendix A: Suppliers, Manufacturers, and Resources 233
Appendix B: Wireless Standards 243
Glossary 249

Index 257

Contents

Acknowledgments ix

Part I: Understanding the Threat   1

Chapter 1: A Brief Overview of Wi-Fi 3

Introducing Wi-Fi 3
    What is wireless networking? 4
    Comparing wireless and Ethernet 5
    How Wi-Fi works 9
Examining Wi-Fi Equipment 11
    Looking at access points 11
    Checking out adapters. 13
Equipment Worth Avoiding 14
    Skipping defunct standards 14
    Passing on nonupgradeable gear 15
Demystifying Wireless Standards 15
    Introducing 802.11: At the base of it all 16
    Explaining 802.11b 16
    Comparing 802.11a 17
    Speeding up with 802.11g 17
    Better security with 802.11i. 17
    Speeding toward the future with 802.11n 18
    Expanding with Bluetooth and 802.15.1 18
Introducing Organizations and Certification 20
    Meeting the Wi-Fi Alliance 20
    Introducing the IEEE 21
Summary. 21

Chapter 2: Network Fundamentals and Security Concerns. 23

Explaining networking terms 23
    Node 24
    Protocol 24
    Network 24
    Internetwork, Internet, intranet 25
    Host 26
    Server 26
    Client. 26

Service. 26
Workstation 27
Peer 27
Examining Network Models 27
Deconstructing the OSI reference model 27
Comparing the TCP/IP model. 29
Examining Wi-Fi layers. 31
Demystifying the TCP/IP protocols 32
Defining important protocols 32
Understanding packet switching and encapsulation 33
Understanding Network Addresses and Names. 35
Explaining IP addresses 35
Demystifying address classes. 35
Exposing restricted addresses 36
Explaining subnets 36
Masking subnets 38
Dynamic host configuration protocol 38
Domain names 39
Root domains 40
Subdomains and host names 40
Connecting Networks. 40
Network bridges 41
Routers and gateways 41
Summary. 43

Chapter 3: The People Behind the Problem 45

Separating Hackers from Crackers 45
Meeting the white hats 46
Avoiding the black hats 47
Understanding the gray hats 47
Identifying script kiddies 47
Hacktivists 49
Visiting the Underground Cracker Culture 50
What Makes Hackers and Crackers Tick 51
Looking for knowledge 51
Fulfilling greed 52
Inflating their egos 52
Pursuing revenge 52
Hacker Conventions and Gatherings 53
DEFCON 53
HOPE 53
ToorCon 54
2600 meetings 54
Worldwide Wardrive 54

        Examining Crackers' Tricks 54
            Social engineering 55
            Trashing 56
        Entertaining Hacking/Cracking Resources 57
        Summary 58

Chapter 4: Hijacking Wi-Fi 59

        Hijacking Sessions 59
            Spoofing 60
            Explaining race conditions 64
            Public hotspots 64
            Protecting yourself at hotspots 65
            Exposing the man in the middle 68
        Understanding Rogue Access Points 68
        Uncovering Denial of Service Attacks 70
            WPA denial of service 72
            Disassociate frame attack 73
            Strong signal jamming 73
            FakeAP flood 74
        Summary 75

Chapter 5: More Wireless Attacks 77

        Attacking Hosts 77
            Using OS weaknesses 79
            Using known security issues 83
        Sniffing the Network 85
            How sniffing works 86
            The promiscuous NIC 87
        Summary 88

Chapter 6: Wardriving 89

        Introducing Wardriving 89
            How wardriving works 91
            Hiding your network 91
            Changing default settings 92
            Avoiding DHCP 94
            Filtering network traffic 95
            Using encryption 95
            Legality of wardriving 96
        Warchalking 97
        Warspying 98
        Software for Defense 99
        More About Wardriving 100
        Summary 100

Chapter 7: Viruses and Wi-Fi 101

Understanding Malicious Software 101
Examining Computer Viruses 104
    Dangerous and double extensions 106
    The Microsoft fink-fund 109
Exposing Trojan Programs 110
Discovering Internet Worms 111
Discovering Blended Threats 114
Bot Software 116
Introducing Spyware 118
Investigating Viruses on Handhelds 120
Understanding the Hoax Problem 120
Summary 122

Part II: Protecting Yourself   123

Chapter 8: Technical Pitfalls and Solutions 125

Discovering Common Wi-Fi Problems 125
    Explaining multipath interference problems 125
    Adjacent channels overlapping 130
    Identifying theft of service 131
    Uncovering configuration errors 132
Detecting and Dealing with Interference 134
    Identifying interference in your home or office 135
    Identifying interference outside your home or office . . 135
    Avoiding physical barriers 136
Defining Interoperability Issues 138
Learning About Antennas 138
    Improper mounting 140
    Grounding antennas 142
    Delineating the Fresnel zone 143
    Interference from trees 144
Optimizing Equipment 145
Summary 146

Chapter 9: Wireless Privacy Concerns 147

Why Privacy Matters 147
Wi-Fi Privacy Threats 149
    Understanding location-based services 149
    Exposing spyware 151
    Avoiding adware and drive-by downloads 152
    Tossing your cookies for greater privacy 154

Demystifying Privacy Policies 156
Other Vulnerable Wireless Technology 158
    Intercepting X10 device signals 158
    Peeping in on Wi-Fi video cameras 161
    Eavesdropping on cordless phones 163
Summary 165

Chapter 10: Encryption and Wi-Fi 167

Introducing Encryption 167
    The crypts 168
    Cipher 168
    Plaintext 168
    Ciphertext 170
    Encryption key 170
    A very brief history of encryption 171
    Secret keys 172
Understanding Modern Encryption Techniques 173
    Failings of secret key cryptography 173
    Understanding public key cryptography 174
    Examining hybrid systems 175
    Digital signatures 176
    Digital certificates 177
    Entropy and key strength 178
Debunking Wired Equivalent Privacy (WEP) 179
    Understanding why WEP fails 179
    Why you should still use WEP 180
Investigating Wi-Fi Protected Access (WPA) 182
Introducing Pretty Good Privacy (PGP) 184
Summary 185

Chapter 11: Securing Your WLAN 187

Understanding WLAN Vulnerabilities 187
Changing Dangerous Default Settings 188
    Rethinking the SSID 188
    Changing passwords and usernames 190
    Editing IP addresses 191
Understanding DHCP 193
    Understanding why DHCP can be a problem 194
    Assigning static IP addresses 195
Filtering Network Traffic 197
    Activating MAC address filtering 199
    Implementing IP address filtering 201

Modifying Broadcast Parameters 202
    Adjusting the power 203
    Turning off SSID broadcast 204
    Setting a minimum connection speed 205
Using Encryption 205
    Using WEP 205
    Installing WPA firmware upgrades 206
Securing Clients and Hosts 207
    Locking down Windows XP 208
    More steps to secure your XP machine 213
    Disabling unnecessary services 214
Summary 215

Chapter 12: Protecting Your Wi-Fi Data 217

Identifying Threats to Wireless Data 217
    Examining hardware failure 218
    Exposing outside threats to data 218
Choosing backup systems 219
    Determining what you need to back up 219
    Deciding how often you will back up 220
    Choosing a type of storage 220
    Selecting media for backups 223
    Things you should consider 224
    Backing up via the Internet 225
    Backup software 226
Protecting Your Hardware 228
    Surge and spike protection 228
    Lightning arrestors and grounding antennas 229
    Preventing problems through proper maintenance 230
Summary 232

Appendix A: Suppliers, Manufacturers, and Resources 233

Wi-Fi Networking Hardware 233
Wi-Fi Resources 236
Security, Privacy, and Antivirus Resources 238
PDA, PC, and Accessory Manufacturers 240
Software 241

Appendix B: Wireless Standards 243

Glossary 249

Index 257

Introduction

Welcome to *Caution! Wireless Networking: Preventing a Data Disaster*. Wireless networking is quickly replacing Ethernet networks in many of our homes and offices. The pace of adoption of Wi-Fi technology has been remarkably quick, primarily because wireless networking equipment is easy to set up and use. In fact, in most cases it's less complicated than setting up an Ethernet network.

Unfortunately, the downside to user-friendly Wi-Fi gear is that the majority of people setting up wireless networks in their homes or offices are not securing them correctly, if at all. Unlike Ethernet networks that require a user to be attached or plugged-into the network, Wi-Fi networks broadcast their signal in all directions, allowing anyone with a wireless adapter to access the network. If you haven't taken the time to secure your Wi-Fi equipment, you may be sharing your files and Internet connection with your neighbors, or worse, hackers.

There are plenty of wireless books available that address basic wireless networking or the setup of specific products. In this book, I focus on the security and safety consequences of using these devices and try to explain these issues in a manner that's helpful to new and intermediate wireless users.

I won't discuss specific brands or provide step-by-step instruction for setting up your access point or router; my goal is to illuminate you about the risks associated with your new wireless equipment and provide some solutions for improving the security on your network. Using this book you'll learn the following things:

- ✦ How TCP/IP networks work
- ✦ How Wi-Fi networks operate
- ✦ Why wireless technology is insecure
- ✦ How you are vulnerable
- ✦ How you can protect your computer and networks

Once you have a basic understanding of the topics covered in this book, I hope you'll be interested enough to learn more about them. Continued education is one of the most important things that you can do to maintain security of your computer and network.

Whom This Book is For

If you're an average computer user — beginning or intermediate — and you've recently installed your own wireless network, then this book is for you. You don't need to be a technical guru to benefit from the information I've presented inside this book. While a general knowledge of computing and familiarity with basic networking and Internet concepts is desirable, Part I provides an intro to networking that will have you up to speed in no time.

This book is for beginning or intermediate computer users who need a better understanding of the security issues surrounding wireless networks, and how to address them on their own network before they become a victim.

How This Book Is Organized

I've divided this book into two parts; each part includes chapters that address a common topic. If you're relatively new to networking or only have a passing familiarity with TCP/IP and wireless security, I suggest you start in Part I and read the book in order. If you're in a hurry or concerned that your network may be compromised, you can skip ahead and return to the earlier material later. Here's how the parts are organized:

- **Part I: Understanding the Threat** — Part I introduces wireless technology and basic TCP/IP networking. It introduces the security and safety problems associated with wireless networks and with networking in general.

- **Part II: Protecting Yourself** — In Part II, I discuss the steps you can take to secure your network, protect your equipment from viruses, protect and recover your data, and maintain your privacy. If you're concerned that your network is not secure, and you already have a grasp of the issues presented in Part I, you can start here.

- **Appendixes** — The appendixes provide useful resources relating to wireless networks. I've also included a glossary of many of the terms I've used in this book.

It's my goal to acquaint you with wireless technology and the related security issues so that you'll be encouraged to pursue more information and improve your computing knowledge. Wireless networks are here to stay, and I want you to get the most out of them without leaving yourself vulnerable.

Special features and icons

At the beginning of each chapter, you'll find a short list of the broad topics covered therein. Throughout the text, you'll encounter icons that I've used to bring different topics to your attention. Here's what each of these icons indicates:

Note Note icons provide important related information about a subject.

On The Web The On the Web icon provides addresses to online resources about a topic.

Caution If I use the Caution icon, I'm warning you about something dangerous where you need to be particularly diligent. I don't use this icon often, but if you see it please pay close attention to the information.

Cross-Reference The Cross-Reference icon directs you to related information elsewhere in the book.

I hope you find this book useful and informative and that it educates you about wireless networking and security and assists you in securing your wireless network.

Part I

Understanding the Threat

In This Part

Chapter 1
A Brief Overview of Wi-Fi

Chapter 2
Network Fundamentals and Security Concerns

Chapter 3
The People Behind the Problem

Chapter 4
Hijacking Wi-Fi

Chapter 5
More Wireless Attacks

Chapter 6
Wardriving

Chapter 7
Viruses and Wi-Fi

A Brief Overview of Wi-Fi

CHAPTER 1

In This Chapter

Introduction to Wi-Fi

Examining Wi-Fi access points and adapters

Learning which equipment you should avoid

Wireless standards explained

Wireless organizations and certifications

There are many wireless technologies available — enough to baffle the most technically sophisticated of us. Many people who have installed wireless equipment in their homes have no understanding of how the technology actually works; some don't even realize that wireless access points and adapters are radio transceivers. It's hard, if not impossible, to secure and defend a technology that you don't understand, this chapter introduces wireless networking, how it works, and in what ways it is different from Ethernet networking.

Introducing Wi-Fi

Wi-Fi, short for *Wireless Fidelity*, is the consumer-friendly name given to the popular 802.11 family of wireless networking protocols. The Wi-Fi Alliance coined the name as a consumer-friendly alternative to 802.11. The Wi-Fi Alliance is a nonprofit industry organization created to promote the adoption and use of the 802.11x protocols.

The Wi-Fi Alliance did not create the 802.11x protocols; it just promotes them. The *Institute of Electrical and Electronics Engineers (IEEE)* created the 802.11x protocols. The IEEE (pronounced eye-triple-e) has created many of the standards used in computing and networking, including the Ethernet standard (802.3) for wired computer networks. I discuss both the Wi-Fi Alliance and the IEEE at the end of this chapter.

Note Throughout this book, I use the term Wi-Fi when referring to 802.11a and 802.11b. I refer to 802.11g as *Wireless-G*. I use the designation *802.11x* interchangeably for all three of the protocols.

Since its introduction in 1997, Wi-Fi has become the dominant wireless networking technology. The success of Wi-Fi is largely due to how easy it is for consumers to set up a *Wireless Local Area Network (WLAN)* in their homes or offices. The majority of Wi-Fi hardware available is very user friendly, and you can usually have it operational in minutes.

What is wireless networking?

Just as the name says, wireless networking is networking without the necessity of running wires through your walls, ceiling, and floors. Wireless networks use radio waves rather than cables to broadcast network traffic and transmit data. If you've ever had to squeeze under a house or through a crawlspace to run Ethernet cable, then you are likely excited by the prospect of avoiding this in the future. (I know I am.)

> **Note** The network terms that are useful to know are *client*, *workstation*, and *server*. A PC is a workstation, and when connected to a network, it's a client. Other network-connected devices, including game consoles, PDAs, and printers are clients, too. Any device connected to a network (that isn't a server) is a client, and the name implies a network connection.
>
> A server is a computer that provides services to network clients. Most home networks don't have or need servers because clients and network devices (routers, hubs, and so on) often provide the same services that a server normally would.

Wi-Fi networks transmit in the unlicensed public 2.4 GHz or 5 GHz radio bands, so you don't need to get a federal license or take a test to use Wi-Fi equipment. You can find Wi-Fi equipment available for three different protocols: 802.11a, 802.11b, and 802.11g. I present more information about HomeRF and other standards later in this chapter.

Because Wi-Fi networks don't require cables, they are more flexible than Ethernet *local area networks* (LANs). You can modify a WLAN with minimal effort. Adding new network clients or resources like printers doesn't require you to install additional cables or hubs. Usually all that you need to do to add a wireless client is install a Wi-Fi network adapter and configure the client to access the network. Mobility is another advantage of Wi-Fi. Unlike Ethernet, you can move freely anywhere within range of your WLAN's signal and remain connected.

While the initial cost may be higher, Wi-Fi's flexibility often makes it less costly to implement and maintain than Ethernet. The amount of time required for installation is minimal compared to Ethernet, and being mobile instead of tethered to a cable is a benefit that also offsets the cost of the equipment.

You can use Wi-Fi to expand an existing Ethernet LAN. Maybe you already have a network but you would like more freedom and mobility. With the addition of an access point you can add new computers to the network simply by installing wireless adapters (see Figure 1-1).

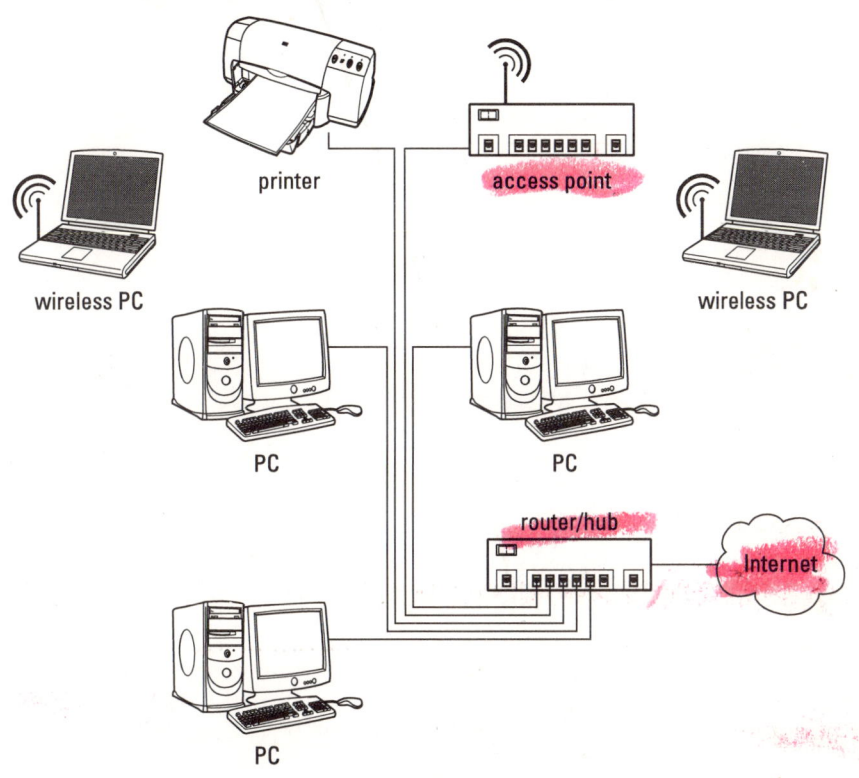

Figure 1-1: A hybrid wireless and Ethernet network

Comparing wireless and Ethernet

Traditionally, the networking of your computers required that you physically connect them by some means, usually with Ethernet cables. Initially Ethernet used coaxial cable, eventually evolving to CAT5 and CAT6 cables. CAT5 cables are usually blue with a large connector that resembles those found on the ends of telephone cords. Most wired home networks use CAT5 cable.

You usually connect computers in an Ethernet LAN together with hubs or routers (see Figure 1-2). In a wired LAN, several Ethernet cables can connect to an Ethernet hub. Network administrators refer to this as a *star* network topology. Hubs can be *passive* devices that merely provide a physical connection between several Ethernet cables, or they can be *active* devices that rebroadcast all of the data they receive in order to strengthen the signal. Manufacturers refer to active hubs as *multiport repeaters*.

Newer hubs have *central processing units* (CPUs) and provide advanced functionality to a LAN environment. Many hubs include routing capabilities. A *router* is a device that directs network traffic on and between LANs. Networks transmit data in *packets*. Each data packet is self contained and addressed to arrive at its destination computer. Routers read the address information in each data packet and route the packet to its destination via the most expedient route.

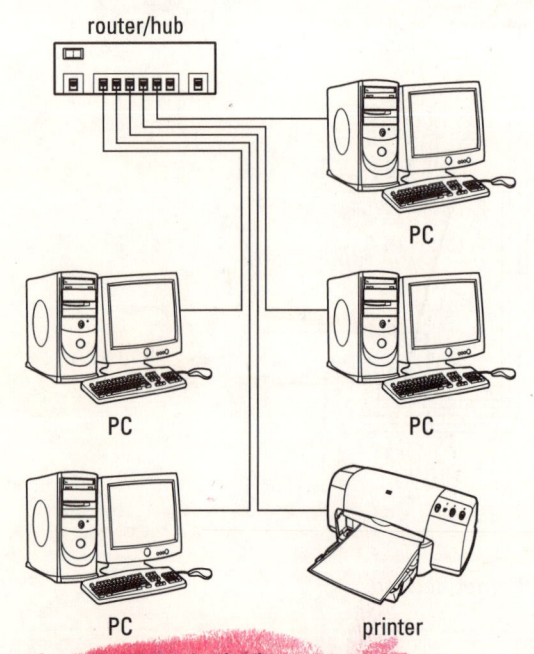

Figure 1-2: A wired Ethernet LAN

A router acts as an interface between two networks; in your home, you may have a router connected to your Internet connection. This router directs traffic between Internet addresses and computers on your LAN. This allows you to have one Internet address and share it among all the computers on your network, giving everyone in the house Web access through the same Internet account.

Ethernet is faster than Wi-Fi (for now) with most home networks theoretically operating at speeds up to 100 *megabits per second (Mbps)*, or millions of bits per second. The fastest Wi-Fi offerings have an advertised speed of 54 Mbps, far slower than Ethernet (see Figure 1-3).

Products are coming to market that utilize various technical tricks to increase throughput, such as transmitting simultaneously on multiple channels, but these are usually proprietary and must be used in a homogeneous network environment; that is, all of the devices must be from the same manufacturer and use the same technology.

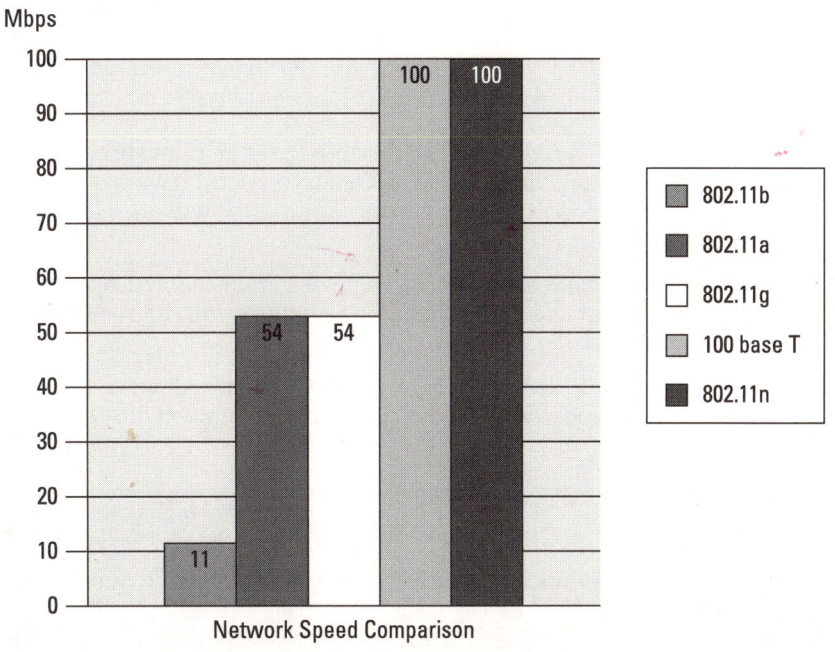

Figure 1-3: Wi-Fi and Ethernet speed comparison

A Wi-Fi network will have less capacity than a LAN; however, unless you typically move huge files across your network, you aren't going to notice a big difference. Your Internet connection also won't be any slower. The speed that you sacrifice for the freedom and flexibility of Wi-Fi is relatively minor, and the newer Wi-Fi gear is making the difference less noticeable.

Even with all of its advantages, wireless is not the solution for everyone. Perhaps you don't mind running Ethernet cables. Maybe you even enjoy it. Why should you bother with Wi-Fi when Ethernet still has plenty of uses and even some advantages? Some reasons to consider Ethernet are:

- **Ethernet equipment is still cheaper than Wi-Fi gear.** With Ethernet, you can network your computers for half the price — sometimes even less — if you can do all the wiring yourself.

Part I ✦ Understanding the Threat

- ✦ **Ethernet is more secure when compared to Wi-Fi.**
- ✦ **Ethernet performance doesn't degrade over distance the same way Wi-Fi does** (but it can degrade over long cable runs).
- ✦ **Ethernet is much faster than current Wi-Fi standards.**
- ✦ **Obstacles and interference don't affect Ethernet the same way they affect Wi-Fi.**

While this may make Ethernet sound good, it's not perfect. Some of the reasons to consider using Wi-Fi are:

- ✦ **Unlike Ethernet, Wi-Fi networks are flexible.** You can relocate computers and equipment with minimum hassle because you don't have to tether them with cables, and you don't have to run any additional network cables to expand your network.
- ✦ **Wi-Fi networks promote mobility.** If you have a Wi-Fi-enabled laptop or PDA, you can move around and stay connected as long as you can receive your WLAN's radio signal.
- ✦ **Wi-Fi is easy to install.** There are no cables to run through walls, crawl spaces, or ceilings, and you don't have to deal with drywall dust or spiders.

The Facts About Network Speed

A company may use the terms data rate and throughput interchangeably when it tries to promote the speed of its networking devices. Data rate actually represents the capacity of a device, though it is sometimes used as if it were the actual speed. If this has left you scratching your head, you're in good company. The abuse and misuse of these terms has led to a lot of confusion, even among IT professionals.

Data rate refers to the amount of data (in bytes) transferred over a network connection in a specified unit of time. For example, an 802.11g device operating at 54 Mbps (millions of bits per second) has the theoretical capacity to deliver 54 Mbps of data. The capacity, or data rate, of any network is not the true measure of its speed.

The throughput of a connection is the truer measure of network speed. *Throughput* is the amount of information sent over a network connection in a specified unit of time. Many factors affect the throughput of a wireless connection, including the number of users on the WLAN, signal interference, signal *attenuation* (weakening), and latency. *Latency* is the amount of time it takes for data to make a round trip between two devices.

The throughput of a WLAN is, without exception, always less than the data rate. Actual throughput may even be less than half the data rate; therefore, users on an 11 Mbps WLAN may have an actual throughput of 6 Mbps or less. Being able to distinguish between these terms will help you make smart hardware-buying decisions.

How Wi-Fi works

In a WLAN, network clients use a *wireless adapter* or *wireless Network Interface Card (NIC)* to connect to the network and communicate with other computers. Each wireless adapter is actually a small *transceiver* (transmitter/receiver) or two-way radio that uses radio waves to transmit and receive network data.

Wireless networks also have devices called *access points* that are stand-alone transceivers that connect wireless clients in a WLAN and act as hubs or routers (see Figure 1-4). Access points often have built-in Ethernet ports, allowing you to connect an access point to a wired LAN that connects the two network segments. Many access points can also connect directly to a broadband Internet connection, making it easy for your wireless clients to connect to the Internet.

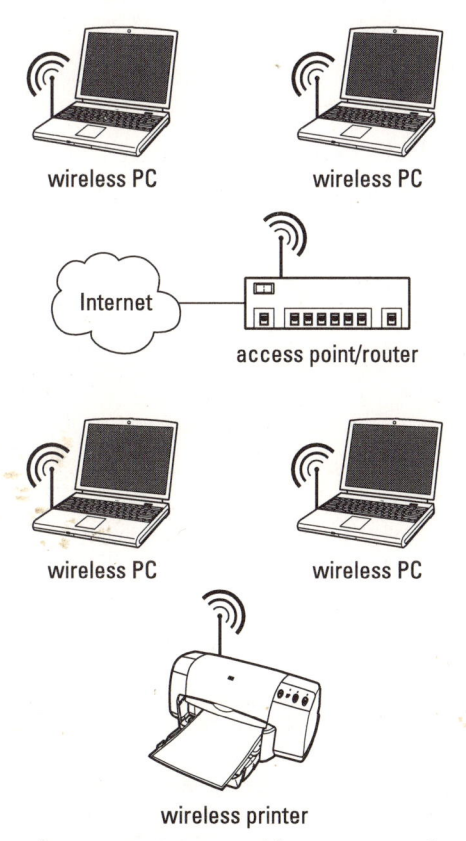

Figure 1-4: A WLAN with an access point and adapters

WLANs can operate in two different modes: *infrastructure mode* and *ad hoc mode*. Infrastructure mode requires the use of access points to connect all of the clients on the WLAN. Ad hoc mode does not use an access point; instead, clients in an ad hoc network communicate directly with each other (see Figure 1-5). Ad hoc mode is slower than infrastructure mode, and I don't recommend it for connecting more than a few wireless clients. However, ad hoc mode is handy for connecting two or three laptops to share files wirelessly. Infrastructure mode is the most commonly used method of connection and the mode that I focus on in this book.

Wi-Fi networks broadcast data through the public airwaves rather than over network cable. The key word to remember is *broadcast*, access points and wireless adapters transmit radio waves in all directions, and anyone within range of your broadcast can receive the signal, and potentially eavesdrop and gain network access.

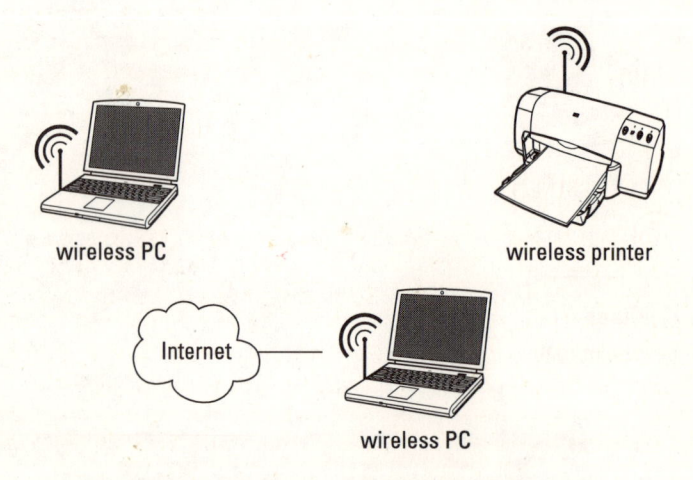

An ad hoc network doesn't use access points.

Figure 1-5: Ad hoc mode

This is the reason that wireless networks are inherently less secure than wired LANs are. To intercept data on a wired LAN, an intruder must have physical access to the network either by connecting over the local Ethernet LAN or through the Internet. However, an intruder doesn't have to connect to a WLAN via a cable. He can sit in an adjacent office or in a parking lot and receive the radio signal.

As long as he can receive the signal, he can record data. Even if you *encrypt* (encode or scramble) all of the data on your WLAN, being able to collect it gives the intruder the opportunity to *decrypt* (decode or unscramble) your data at his leisure. The current methods of encrypting Wi-Fi data aren't especially robust and an intruder can defeat them, possibly compromising your network and data.

Cross-Reference: I discuss encryption in Chapter 10 and attacks against wireless encryption standards in Chapter 5.

Security firms, antivirus makers, and firewall vendors all like to exaggerate the threats to wireless networks, and to some extent, they are genuine, but not nearly as grave as the experts make it sound.

Cross-Reference: You can find out how to sufficiently secure any network in Chapter 11.

Examining Wi-Fi Equipment

When you're setting up a home or small office WLAN, there are primarily two types of hardware that you'll encounter: access points and wireless adapters. There are many different types of adapters and access points available, each with different features, advantages, and disadvantages. Here are a few of the types and configurations that you will encounter.

Looking at access points

The central piece of Wi-Fi hardware in your WLAN is the access point. It sends and receives signals between clients on the network and provides a central point of connectivity for your WLAN. Every WLAN needs at least one access point in order to operate in infrastructure mode. In addition to providing connectivity to wireless NICs, an access point can also interface with an Ethernet network or broadband Internet connection. Some of the additional services that an access point may provide are:

- **Dynamic Host Configuration Protocol (DHCP) services** — DHCP eliminates the need for you to configure each computer with a static (unchanging) Internet Protocol (IP) address. The DHCP service automatically assigns IP addresses to clients as they connect to the WLAN.

- **Network switch, or hub** — These connect wired PCs to the network using Ethernet cables.

- **Router services** — A router allows multiple users to share a single broadband Internet connection and directs network traffic by routing data between clients.

- **Print server** — A few access points have one or more places to connect printers (both serial and USB ports), so that you can share the printers on your WLAN. You can also purchase Wi-Fi adapters for your printers and eliminate the need for this function in your access point.

Some access points can operate in one or more modes: *normal*, *client*, *bridge*, and *repeater* (see Figure 1-6). The definition of each of these modes follows:

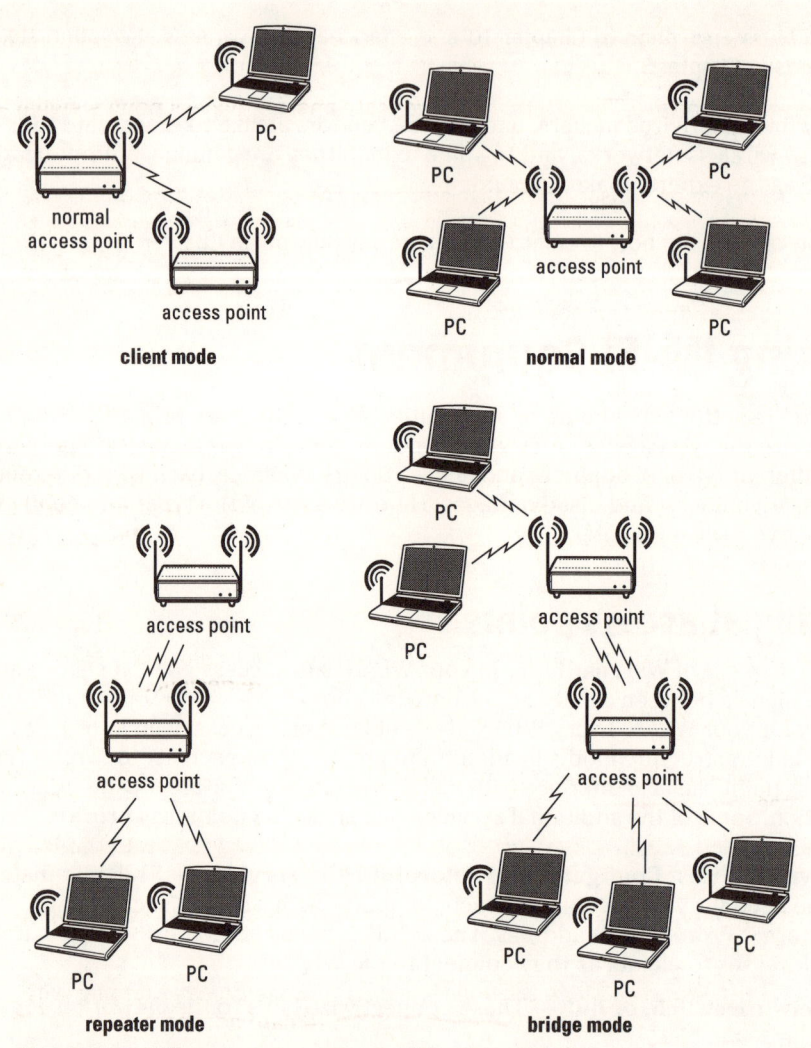

Figure 1-6: Different access point modes of operation

- **Normal mode** — The access point operates normally and provides a central point of connection for clients.
- **Client mode** — The access point operates as a network client (like a wireless adapter) and only communicates with other access points, not clients.
- **Bridge mode** — When acting as a bridge, an access point communicates directly with another access point. Both access points must be capable of

point-to-point bridging and usually have to be from the same manufacturer. A network bridge is useful for extending a WLAN between buildings.

- **Repeater mode** — The access point repeats another access point's signal and extends its range.

You have to select an access point that supports the same standard as the wireless clients on your network. If you fail to do this, the access point will be unable to communicate with clients that use a different standard. You can purchase a dual-mode access point that supports two or more standards, but be prepared to pay a bit more for this added functionality.

Checking out adapters

Wireless network interface adapters translate between your computer and the network and allow your computer to speak the language of the network and communicate with other clients. They perform the same function on a Wi-Fi network that Ethernet cards do on a wired LAN. Frequently called a NIC, a wireless adapter is essentially a two-way radio that can send and receive radio signals.

Each client has to have a wireless adapter in order to connect to the WLAN and communicate with other computers. Wireless adapters are either internal or external, depending on the client device for which they are designed (see Figure 1-7). Types of adapters include:

- **USB wireless adapters** — These adapters connect to the USB port on a client device.
- **PCI adapter cards** — These are internal adapters that insert into the PCI slot in a PC.
- **PCMCIA card adapters** — These adapters connect via the PCMCIA slot on a notebook.
- **Built-in adapters** — Many new notebook computers come equipped with internal wireless adapters, rather than PCMCIA card adapters.
- **Compact Flash (CF) card adapters** — These adapters fit in the CF slot on PDAs and other handheld devices.

When you're selecting an adapter, make sure that it's compatible with your access point and the device for which it's intended. There are adapters available for just about every type of client device, including computers, printers, and game consoles. Check the device's documentation or the manufacturer's Web site to find out which adapters are compatible with it.

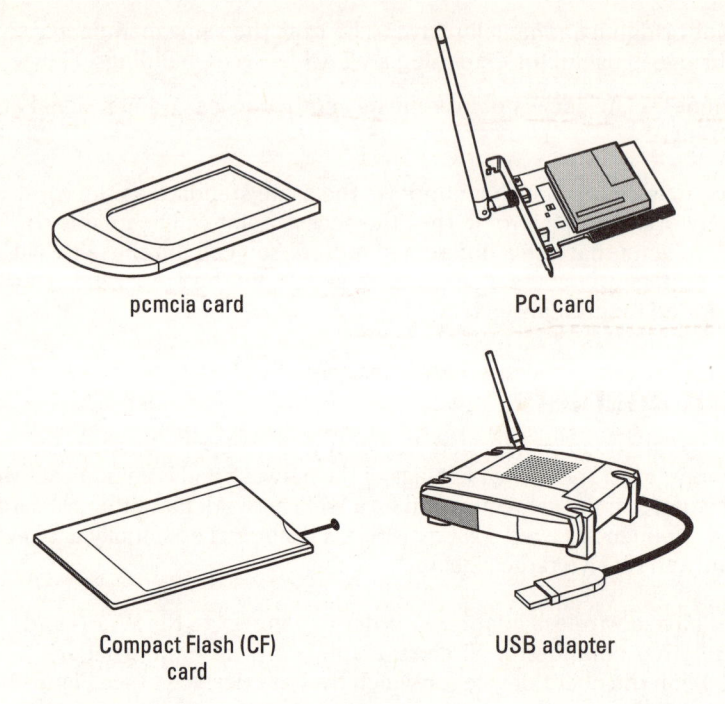

Figure 1-7: Types of wireless adapters

Equipment Worth Avoiding

Not all wireless networking gear is equal. There are defunct standards and non-upgradeable devices that you'll want to avoid altogether because they create problems.

Skipping defunct standards

Wi-Fi wasn't always the only major player in the home wireless networking market. Around the same time that the IEEE introduced the 802.11b standard, another WLAN technology, HomeRF, also entered the marketplace. HomeRF operated in the same 2.4 GHz radio frequency band that 802.11b does. HomeRF was slow, about 1 Mbps, and extremely insecure compared to Wi-Fi.

The HomeRF working group supported and developed the HomeRF standard. The HomeRF working group disbanded when it became clear that its standard was no competition for 802.11b. HomeRF is a dead technology with no support or further development. While it's unlikely that you'll find HomeRF gear for sale in a retail store, the equipment is often available on auction sites. Unscrupulous vendors try to dump useless HomeRF equipment, pitching it as the latest and greatest thing. When shopping for Wi-Fi equipment, be sure that you purchase 802.11x gear and avoid HomeRF if you see it for sale.

The easiest way for you to ensure that you are purchasing Wi-Fi-compliant equipment is to look for the "Wi-Fi certified" logo on the package. This certification is part of the Wi-Fi Alliance's effort to ensure the interoperability of 802.11x equipment.

Passing on nonupgradeable gear

802.11b has been around since the late 1990s, and there is a lot of equipment available based on this standard. You can find bargains if you look for older 802.11x gear, but bargain hunting can backfire. Much of the older equipment isn't upgradeable, so as the IEEE and manufacturers introduce improvements to the 802.11x protocols, you won't be able to take advantage of them.

The *firmware* on many access points and adapters is upgradeable. The EPROM and EEPROM chips holding the programming code that operates these devices continue to hold their programming even when they have no power. Professionals call this firmware because it falls between the definitions for *hardware* and *software*. A user, using executable upgrade programs supplied by the device's manufacturer, can replace the programming code on firmware chips.

Manufacturers use firmware upgrades to introduce improvements or new capabilities to a device. Most recently, manufacturers have released security upgrades based upon the 802.11i standard. (802.11i is the standard that describes improved security for Wi-Fi devices.) The Wi-Fi Alliance based the new *Wi-Fi Protected Access (WPA)* encryption standard on portions of 802.11i.

Being able to take advantage of firmware upgrades is an important thing to consider when you are purchasing new equipment. You can find out if a device has upgradeable firmware by checking the manufacturer's Web site or, when possible, the device's documentation.

Demystifying Wireless Standards

In the past few years, the marketplace for WLAN products has grown at an incredible rate, with organizations introducing new standards and improving existing ones. This hasn't made wireless networking any less confusing for the average consumer. The Wi-Fi Alliance adopted the term Wi-Fi because it is less confusing than the series of numbers used to define the various wireless protocols. Still, it's necessary to understand the principal wireless protocols to avoid confusion and costly mistakes.

For wireless devices to operate on the same network, they must use compatible standards and operate on the same frequency. For example, even though 802.11b and 802.11a are both Wi-Fi standards, they can't communicate or interoperate because these standards don't use the same frequency band or technology.

> **Note** Wi-Fi devices operate only with compatible devices that use the same frequency band (such as 2.4 GHz or 5 GHz), but there are certified *dual-band* products that can operate using both frequencies. These dual-band devices contain both 802.11a and 802.11b transceivers and can communicate with devices using either standard.

Introducing 802.11: At the base of it all

First introduced in 1997, 802.11 consists of the family of wireless standards developed by the IEEE. Currently three primary *physical layer standards* exist within 802.11. Physical layer standards describe the network medium and the mode of transmission; in the case of Wi-Fi the physical standards describe the frequency band, and the transmission technology used to access and communicate on the network operating in that band.

The physical layer standards are 802.11a, 802.11b, and 802.11g. When you're shopping for Wi-Fi gear, these are the standards that you'll encounter. Other standards, like 802.11i, describe different aspects of wireless networking, and you don't really need to know them unless you're an IT professional.

Explaining 802.11b

802.11b was the first 802.11x protocol introduced, actually appearing prior to 802.11a (the IEEE didn't release the standards in alphabetical order). Wi-Fi has risen to dominate the home wireless market, and because 802.11b has been around the longest, there is an abundance of 802.11b gear available. This means that 802.11b equipment is inexpensive compared to 802.11a and Wireless-G devices.

802.11b devices operate in the 2.4GHz radio band, with a maximum capacity of 11Mbps. Heavy traffic on the same channel can significantly reduce throughput, and speed can decrease the farther you get from an access point. 802.11b divides the 2.4GHz network into 11 channels, although devices in a network usually utilize three in order to limit the chance of access points interfering with one another.

Other consumer electronic devices, including cordless phones and microwave ovens, also use the 2.4GHz band. When operating, these devices may interfere with 802.11b WLAN functions. If you already own other 2.4GHz devices, you may want to consider this before deciding which wireless standard to use.

> **Cross-Reference** I discuss ways of dealing with *radio frequency interference (RFI)* in Chapter 8.

Comparing 802.11a

802.11a operates in the 5GHz band and has a maximum capacity of 54 Mbps, almost five times faster than the maximum capacity of 802.11b. Realistically, 802.11a must be close to an access point to achieve maximum throughput. Compared to 802.11b and Wireless-G, the range of an 802.11a access point is significantly shorter. Because of this, 802.11a isn't the best choice for providing Wi-Fi access to a wide area, unless you are willing to invest in multiple access points so that the signal reaches all clients.

802.11a devices aren't compatible with 802.11b devices, because they operate on different frequencies and use different technologies. If you have an existing 802.11b network, you'll have to purchase new equipment for every network client.

Speeding up with 802.11g

802.11g, or Wireless-G, is the newest 802.11x physical layer standard. 802.11g has the same maximum throughput as 802.11a, but operates in the 2.4GHz band along with 802.11b. Wireless-G is backward compatible with 802.11b, and devices for both standards can interoperate on the same wireless network. An 802.11g access point can communicate with 802.11b cards, so you can upgrade your clients incrementally; but to take advantage of full Wireless-G throughput, all of your clients and access points must be Wireless-G devices.

Note Although Wireless-G is backward compatible with 802.11b, you'll notice slower WLAN throughput on a network with mixed clients and access points (Wireless-G and 802.11b) than in a homogeneous network built only with Wireless-G gear.

Better security with 802.11i

Wired Equivalent Privacy (WEP) is the original Wi-Fi encryption algorithm used to secure communication on 802.11x networks. Experts defeated WEP, making it useless (well almost, but not entirely). In response, the IEEE began to develop 802.11i, the next Wi-Fi security standard. 802.11i addresses security concerns in Wi-Fi and improves encryption, user authentication, key management, and distribution.

Meanwhile, the Wi-Fi Alliance has developed and introduced Wi-Fi Protected Access (WPA) as a replacement for WEP. The Wi-Fi Alliance based WPA on a subset of the 802.11i standard, and WPA provides better security than WEP. If possible, you should upgrade your gear to use WPA.

Note Although WPA is a great improvement over WEP, it isn't perfect. There are ways that an intruder can defeat WPA, but these are more the result of poor password and passphrase selection by users and not flaws in WPA itself. Refer to Chapter 10 for WPA best practices.

> ### Wireless-G versus 802.11a
>
> Because it offers throughput comparable to 802.11a and is backward compatible with 802.11b, Wireless-G can be a good alternative to 802.11a. Wireless-G has price and performance advantages over 802.11a, especially in a small home or office network; but when system capacity is important, 802.11a has the advantage. 802.11a has more channels available and can support more traffic, which is why it's a good choice in many enterprise environments. For mission-critical applications, 802.11a is a better choice.

Speeding toward the future with 802.11n

IEEE has recognized the 802.11n working group, which has begun developing the next 802.11x physical layer standard. The IEEE should introduce 802.11n by 2006 and expects it to provide up to 100 Mbps of actual throughput, not just 100 Mbps data rate. For example, the data rate for Wireless-G is up to 54 Mbps, but the actual throughput is often less than half of that.

Expanding with Bluetooth and 802.15.1

Bluetooth is a *wireless personal area network* (WPAN) technology developed by the Bluetooth Special Interest Group, an industry organization founded by Nokia, Ericsson, IBM, Intel, and Toshiba. Developers named Bluetooth after a twelfth-century king, Harald Blatand (Bluetooth), who unified Denmark and Norway. The goal of Bluetooth technology is to enable users to connect many different devices simply and easily without cables.

Although Bluetooth operates in the 2.4 GHz band, it doesn't compete directly with Wi-Fi because it is too slow (1 Mbps) and its signal range is too short for a WLAN. Bluetooth devices coexist peacefully and occasionally interoperate with Wi-Fi networks.

Bluetooth connects devices and peripherals without annoying cables. You can already purchase keyboards, mice, printers, digital cameras, and PDAs that employ Bluetooth technology. Bluetooth has even found its way into some automobiles as a way to connect digital music players to the car's stereo system.

Bluetooth devices can connect and create small ad hoc networks called *piconets*. In each piconet, the device that first initiates the connection becomes the *master* device. Each master device in a piconet can manage up to seven *slave* devices. Figure 1-8 illustrates a master and six slaves, leaving room for one additional device.

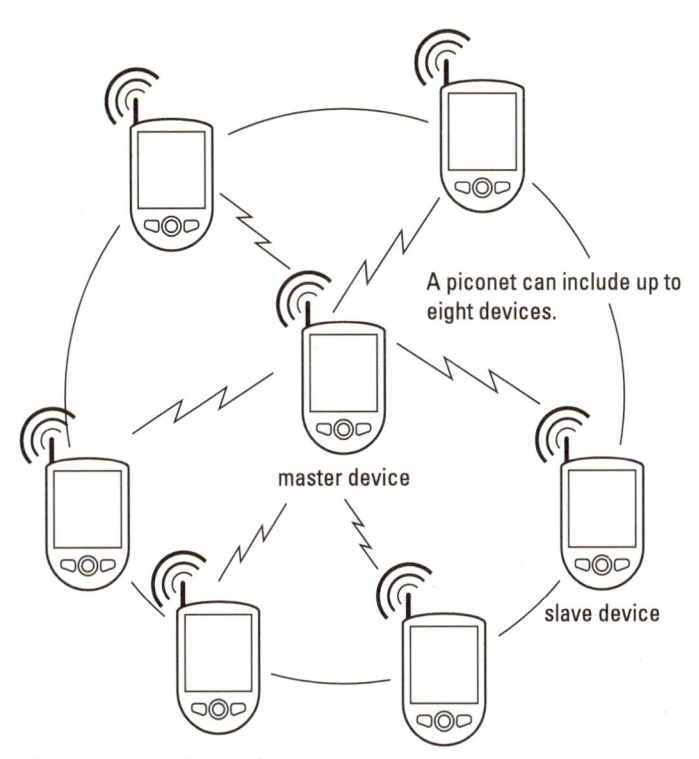

Figure 1-8: A Bluetooth piconet

A device can be a member of multiple piconets at one time, but can only be a master in one. *Scatternets* are piconets connected by one or more devices (see Figure 1-9). A scatternet may contain 10 fully loaded piconets at any one time.

Bluetooth is a useful and complementary technology to Wi-Fi, and you will begin to see it integrated into more devices every year. You can upgrade your printers and some other peripherals to use Bluetooth by installing small Bluetooth adapters. These adapters are inexpensive, with some costing less than $20.

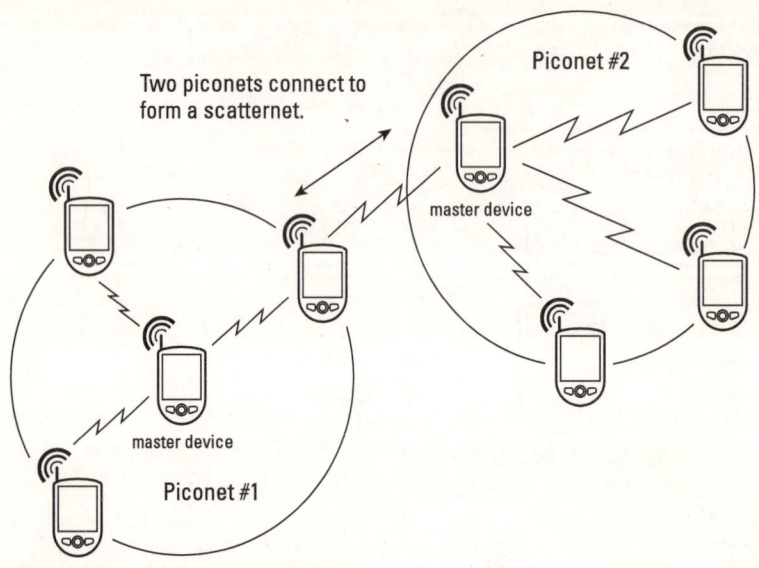

Figure 1-9: A Bluetooth scatternet

Introducing Organizations and Certification

Two primary organizations develop or promote wireless standards. The first, the Institute of Electrical and Electronics Engineers (IEEE) is an international, technical, and professional organization that exists to define and promote standards among engineers.

The second organization, the Wi-Fi Alliance, is a nonprofit industry organization with more than 200 member companies that certify wireless product compliance to the 802.11x standards. Where the IEEE exists to develop standards, the mission of the Wi-Fi Alliance is to promote these standards to the public through product testing and certification.

Meeting the Wi-Fi Alliance

Through its certification programs, the Wi-Fi Alliance encourages manufacturers to comply with 802.11x standards when designing wireless devices. The Wi-Fi Alliance ensures that devices are 802.11x compliant by having them independently tested. The Wi-Fi Alliance also conducts tests to ensure interoperability among Wi-Fi-certified products.

You can be sure that a network component will interoperate with other devices that use a compatible standard if it carries the *Wi-Fi certified* logo. This allows you to use devices from different manufacturers with confidence that they will work together.

Introducing the IEEE

The IEEE (sometimes pronounced "eye-triple-ee") uses consensus-based working groups to develop standards. Manufacturers use IEEE standards as blueprints for developing compatible products, solutions, or processes. IEEE standards give us the opportunity to choose among different vendors for compatible wireless equipment. Without the IEEE as a standards body, manufacturers would develop multiple incompatible standards, instead of adhering to common ones.

> **Note**
>
> Many manufacturers still define their own standards rather than support and promote common ones. Usually this is an attempt to monopolize a specific market area and lock in a customer base dependent on their products and support.
>
> Avoid proprietary technologies when possible; they are usually more expensive and you'll be at the mercy of a single manufacturer, which could be especially bad if it later abandons its own standard.

Summary

In this chapter, I introduced the technologies and standards common to wireless networking. The subjects I presented include:

- ✦ An introduction to wireless networking or Wi-Fi
- ✦ Wi-Fi access points and wireless adapters
- ✦ Which equipment you should avoid, and why
- ✦ An explanation of the common wireless standards you'll most likely encounter
- ✦ The chief wireless organizations and certifications

This information will help you to understand these topics when I discuss them in more depth in later chapters, and aid you in choosing equipment when planning a new WLAN or extending an existing one.

✦ ✦ ✦

Network Fundamentals and Security Concerns

CHAPTER 2

In This Chapter

Introducing networking

Network models explained

Introduction to TCP/IP

Network addressing and names

Understanding bridges, routers, and gateways

Networking can be a confusing subject, but it doesn't have to be. The material in this chapter will make it easier for you to understand some of the threats that I introduce in later chapters. It can also help you when it comes time to troubleshoot some of the technical pitfalls you'll likely encounter when using wireless networking devices.

This chapter introduces some networking terms and technology, and I attempt to explain, clearly and concisely, how much of it works. Don't panic. The techno-babble is minimal and the information that is presented is limited to what you need in order to get the most out of the material in subsequent chapters.

Explaining Networking Terms

Throughout this and any other book about networking computers, you are going to encounter network terminology. In order for you to get the most out of the information in this book, I'm going to start by defining a few of the most common and important terms that you are going to see. I mentioned a couple of these terms in Chapter 1; now I'll present them in more detail.

Node

Put simply, a *node* is a network junction or point of connection. Any device connected to a network that can send and receive data is a node. Your computer is a network node; if you have a broadband Internet *router*, that's a node, too. Your access point is a node on your wireless local area network (WLAN), as is an Ethernet hub or switch.

> **Note** A router is a hardware device used to interconnect networks, such as your WLAN and the Internet, and route data traffic between them. I discuss routers, and the related terms *bridge* and *gateway,* later in this chapter.

Protocol

The word *protocol* appears often in this book. I talk about the 802.11x protocols, TCP/IP suite protocols, and more. A protocol is a set of instructions that describes how nodes on a network communicate. Protocols determine how nodes send and receive data.

There are two types of protocol: *open* and *proprietary.* Open protocols are developed and supported by most organizations and manufacturers. Proprietary protocols are the property of one organization or company that licenses the technology to third parties.

Open protocols are usually better supported, and allow different manufacturers to produce compatible equipment at a lower cost to the consumer than is possible using proprietary technology. The main protocols discussed in this book are all open protocols.

802.11x protocols describe how Wi-Fi nodes communicate, and TCP/IP describes how nodes communicate on the Internet and other networks. Protocols allow different operating systems (Windows XP, MAC OS X, Linux, etc.) to share information and communicate by giving them a common language or way to communicate since each operating system is proprietary technology, not usually designed to interoperate with other systems.

Network

For our purposes, a *network* is any arrangement of interconnected nodes that can share data. There are several different physical designs or *topologies* of data networks (described later in this chapter). Nodes on a network share data using a shared medium (Ethernet, Wi-Fi) and a common language or protocol.

Internetwork, Internet, intranet

An *internetwork* is two or more connected networks or a large network comprised of smaller interconnected networks. You may be thinking, "Isn't that what the Internet is?" Well, yes and no. There was a time when people used the terms interchangeably, and *internet* (not capitalized) was the abbreviated form of internetwork.

Now, *Internet* (capitalized) is used to refer to the global connection of networks that communicate using open protocols and standards, rather than proprietary network technologies. To refer to interconnected networks, including private networks in companies that may operate a proprietary networking technology, use the term *intranet*. Figure 2-1 illustrates three private networks (A, B, and C) or intranets connected with routers. Routers are devices that connect networks and route traffic between them.

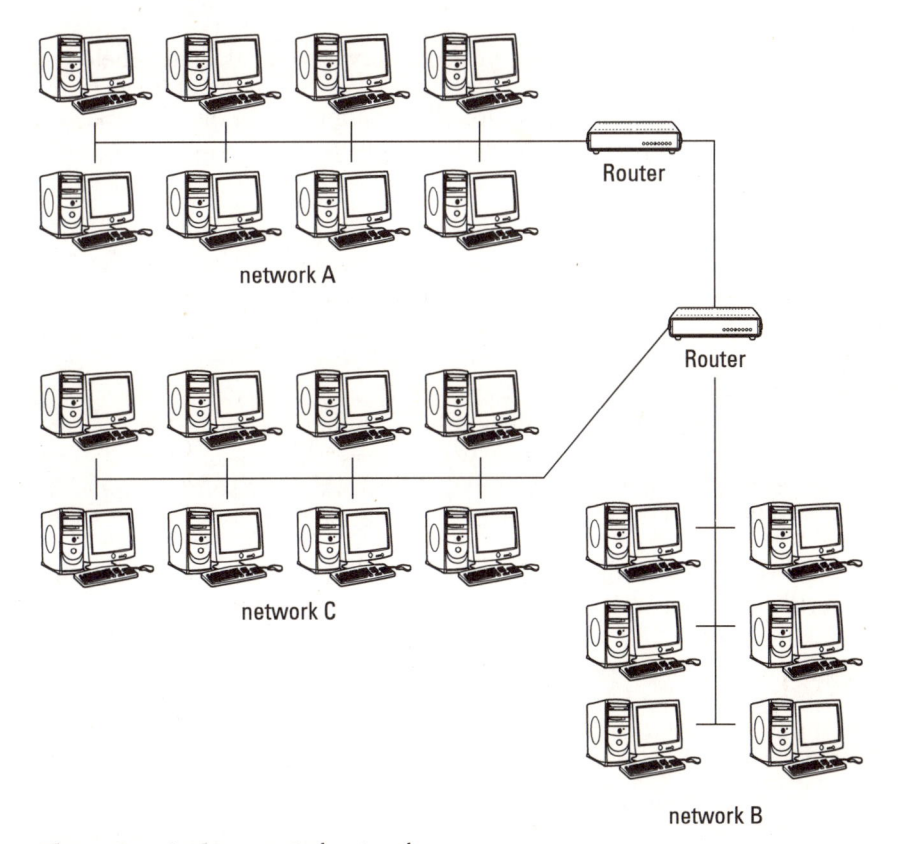

Figure 2-1: Interconnected networks

This can be confusing, because you can connect an intranet to the Internet. An easy way to keep them straight is to remember that the Internet is open and public, while an intranet is private and internal. The latter is the network within your home or company, while the former is the network beyond your walls that we all share.

Host

A *host* is a type of node on a network that shares information and sometimes services. You can apply the name host to almost any type of network node. A client PC is a host, as is a server (see definitions below), and your PC is a host both to users and to its peripherals (printer, scanner, and so on). Because it's applied to so many different nodes, the term host can be confusing.

Server

On a network, a *server* is a node that provides services (see definition below) to other nodes. A server can be a stand-alone device or computer, or it can be a client machine that shares services with other computers on the network. To confuse things, a server can also be a client when it uses services provided by another server.

The term server can refer to the machine itself as well as to individual applications that it runs (or hosts). Some server applications include Web servers, File Transfer Protocol (FTP) servers, and print servers.

Client

A *client* is a network node that receives services from a server. Your PC is a client; it receives service from e-mail servers, Web servers, and others. By itself, the term client implies that a device is connected to a network. Some software applications can also be clients receiving services and information from server applications.

For example, your e-mail software is an e-mail client that receives service from an e-mail server, and your Web browser is a Web client served by Web servers. A client can also act as a server for other clients by sharing resources or providing services to them.

Service

Network *services* are information or resources provided to clients by a server. Services include printing, messaging, e-mail, network connectivity, and Web access.

Workstation

A *workstation* can be one of two things. When applied to a network, a workstation is just another name for a client PC. The PC with which you connect to the network is your workstation. The term workstation can also describe a very powerful, high-end stand-alone PC. Usually workstations run some variety of Unix, and vendors of popular workstations include Sun, HP, IBM, and Silicon Graphics.

Peer

You're probably starting to hear this term a lot, usually in reference to peer-to-peer networks, such as Kazaa. A *peer* is simply a node on a network that is both a server and a client to all of the other nodes on the network. Because all of the nodes are equal, they are called peers. Peer-to-peer networks are decentralized, with services being spread across all of the nodes.

Examining Network Models

Network models are just abstract ways for engineers to describe how all of the protocols, hardware, and physical media of a network interact. The layered structure of models provides a simple visual reference for what's going on at all levels of a network.

Unless you're going to become an engineer, or intend to develop network software, you really don't need to kill yourself trying to learn network models. In fact, in my experience most professionals have little more than a trivial understanding of the various reference models; they just don't need it.

Keeping this in mind, I'm presenting you with a brief and simplified explanation of three network models. This is just so that you'll be able to grasp related terms and have a better understanding of what different network components do.

Deconstructing the OSI reference model

The *Open Systems Interconnect* (OSI) reference model is the most widely used model for describing communications across networks. In practice, most software and hardware vendors never implemented the OSI model. It was never widely supported, and engineers mostly use it as a teaching tool. If you ever take a networking class or information systems course, you will learn this model.

The OSI model describes network communications using seven layers (see Figure 2-2). These layers are (in reverse order):

Application Layer (layer 7)

Presentation Layer (layer 6)

Session Layer (layer 5)

Transport Layer (layer 4)

Network Layer (layer 3)

Data Link (MAC) Layer (layer 2)

Physical Layer (layer 1)

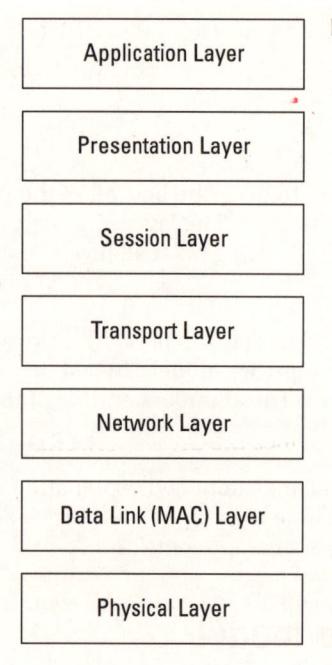

Figure 2-2: The OSI reference model

Each of the layers assigns functions relating to network communications. I present the layers top-down with the top *application layer* (layer 7) being that which is closest to the user, and the bottom *physical layer* (layer 1) being where the actual transmission of electrical impulses takes place. A brief description of each layer follows.

Application layer

As a user, you interact with this layer. Some of the functions that reside at this layer are opening, closing, checking e-mail, and deleting files.

Presentation layer
The presentation layer manages the way systems encode data. Encryption and decryption of transmitted data occur at this layer.

Session layer
This layer describes managing communications between nodes. In real-world systems (remember, OSI is primarily a teaching model), most of this occurs at the transport layer.

Transport layer
This layer describes the processes for managing communications and providing for the integrity of transmitted data. Riveting stuff. Are you still awake?

Network layer
This layer describes the routing of communications. On a TCP/IP network, a router operates on this layer.

Data link layer
This layer is responsible for maintaining the validity and integrity of communication between nodes on a network. Transmitted data may pass through many nodes before reaching its destination. At this layer, each node checks that the data is intact and valid before passing it on to the next node.

Physical layer
This is where the physical operation of the network occurs, and devices pass around data in the form of electrons. The physical layer describes the actual network, whether it's Ethernet, fiber-optic, or, in the case of a WLAN, radio frequencies.

Comparing the TCP/IP model

In contrast with the OSI model, engineers actually use the TCP/IP model, which uses the TCP/IP suite described later in this chapter (see Figure 2-3). It has four layers — compared to the seven in the OSI model — that sit on the physical layer (not described by the TCP/IP model). The four layers of the TCP/IP model are (again, in reverse order):

Application Layer (layer 4)

Transport Layer (layer 3)

Network Layer (layer 3)

Data Link (MAC) Layer (layer 1)

Figure 2-3: The TCP/IP reference model

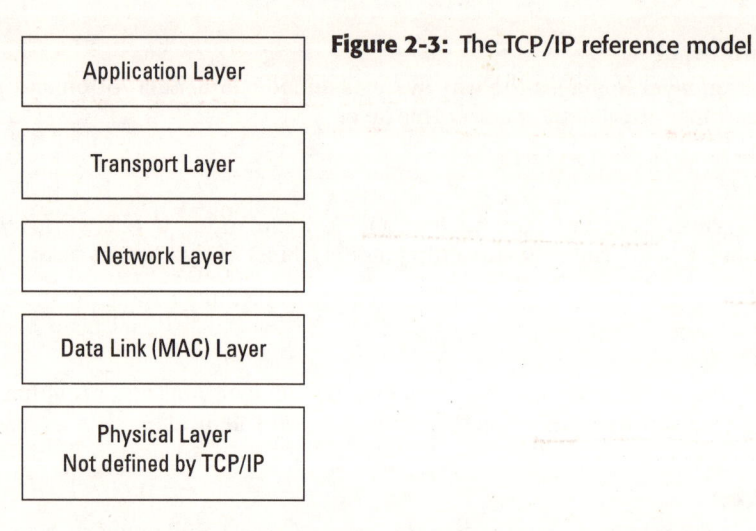

```
┌─────────────────────────┐
│    Application Layer    │
└─────────────────────────┘
┌─────────────────────────┐
│    Transport Layer      │
└─────────────────────────┘
┌─────────────────────────┐
│    Network Layer        │
└─────────────────────────┘
┌─────────────────────────┐
│  Data Link (MAC) Layer  │
└─────────────────────────┘
┌─────────────────────────┐
│    Physical Layer       │
│  Not defined by TCP/IP  │
└─────────────────────────┘
```

Like the OSI model, each of the TCP/IP layers assigns functions relating to network communications. A brief description of each layer follows:

Application layer
Just like the OSI model, this is where the user interacts with network programs. TCP/IP client/server applications communicate on this layer.

Transport layer
This layer uses two protocols — Transmission Control Protocol (TCP) and the User Datagram Protocol (UDP) — to manage communication between hosts. TCP ensures that data arrives at its destination without errors or missing packets, UDP does not provide error checking and is used for maintaining connections between networked applications.

Network layer
This layer uses Internet Protocol (IP) to route data across connected networks.

Data link (MAC) layer
This layer describes the transmission of data across a local network. This occurs once the network layer has routed data across interconnected networks, such as the Internet, and it has arrived at its destination network. This layer also describes transmission of data that never leaves the local network. In the TCP/IP model, engineers often refer to this layer as the *network interface* layer or the *media access control* (MAC) layer.

The TCP/IP model describes how to route data across networks without worrying what the actual physical network is.

Examining Wi-Fi layers

The 802.11x standards define the layers upon which communication takes place on a WLAN. 802.11 defines three layers of the WLAN network model: network infrastructure, media access control, and physical (see Figure 2-4). The layers of a WLAN define how a WLAN operates. Higher layers defined by the TCP/IP suite handle the routing and delivery of data. To oversimplify it a bit, the TCP/IP model just described runs on top of the WLAN model.

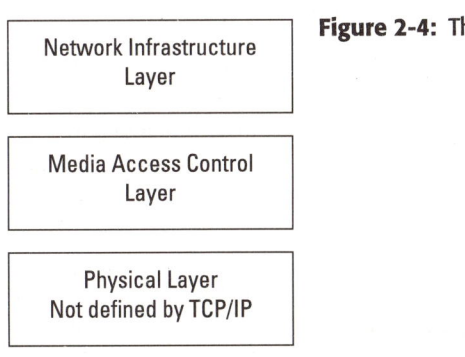

Figure 2-4: The WLAN Layers

- Network Infrastructure Layer
- Media Access Control Layer
- Physical Layer — Not defined by TCP/IP

Network infrastructure layer
This layer describes how devices communicate on a WLAN, and how the network authenticates users and determines who is an authorized user.

Media access control layer
This layer describes how computers on the network gain access to receive and transmit data and how they maintain communication between nodes after authentication has succeeded.

Physical layer
The physical layer includes all hardware components and wireless connections and describes the electrical interface to the network. It determines how nodes transmit and receive streams of data on a radio frequency channel.

The physical layers of a WLAN are all you need to remember. When shopping for equipment, you'll be selecting Wi-Fi equipment based on one of the three 802.11x physical standards: 802.11b, 802.11a, or 802.11g.

Demystifying the TCP/IP Protocols

The TCP/IP suite is the global standard for Internet communications. TCP/IP isn't a single protocol, but a suite containing dozens of protocols. The name refers to two of the most important protocols in the suite, the Transmission Control Protocol (TCP) and the Internet Protocol (IP). IP provides addresses that make TCP/IP communications routable. This means that all messages transmitted over a TCP/IP network are marked with the IP address of the destination network and station.

TCP is one of the main transport protocols in the suite (the other being UDP). TCP ensures that data arrives at its destination without errors or missing packets.

Defining important protocols

The TCP/IP suite contains dozens of protocols. Because network services and devices use some of them more than others, you are more likely to encounter them whenever you read about networking. Table 2-1 lists some of the more important protocols in the TCP/IP suite, but it's by no means complete.

Table 2-1
Common Protocols in the TCP/IP Suite

Protocol Name	Function
ARP	**Address Resolution Protocol.** This protocol matches hardware (MAC) addresses to IP addresses. This protocol is targeted in some of the more technical attacks on Wi-Fi.
DHCP	**Dynamic Host Configuration Protocol.** Remote configuration of network addresses on client machines (more detailed description later in this chapter).
DNS	**Domain Name System.** References host names to IP addresses.
FTP	**File Transfer Protocol.** Describes transferring files between nodes.
HTTP	**Hypertext Transfer Protocol.** Used to deliver and read Web pages.
ICMP	**Internet Control Message Protocol.** Delivers routing control messages.
UDP	**User Datagram Protocol.** Connects applications between network hosts.

Understanding packet switching and encapsulation

All TCP/IP networks, even wireless ones, use a routing technology called *packet switching*. On a packet-switched network, computers divide data into smaller, individually addressed packets. Because each of these data packets contains a destination address, they can each follow different paths to reach their destination. Contrast this with a circuit-switched network, such as the traditional telephone system, where communications require a dedicated point-to-point connection (see Figure 2-5).

Circuit switching creates a dedicated circuit between network hosts for the duration of the session.

Packet switching doesn't require a dedicated circuit. Routers direct data to the correct destination.

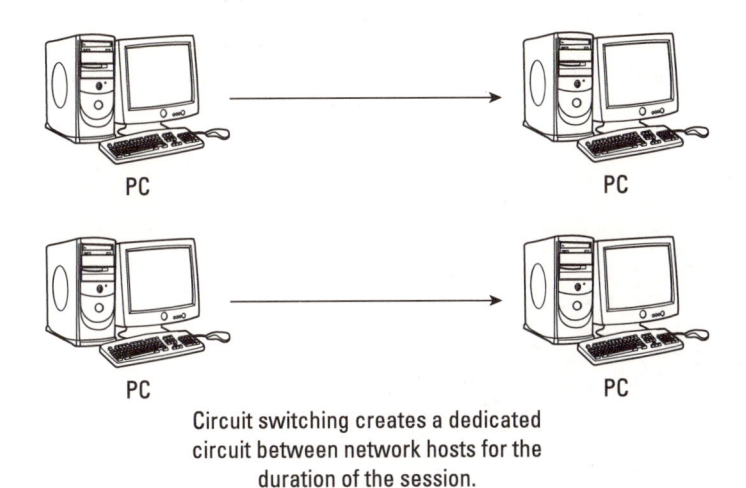

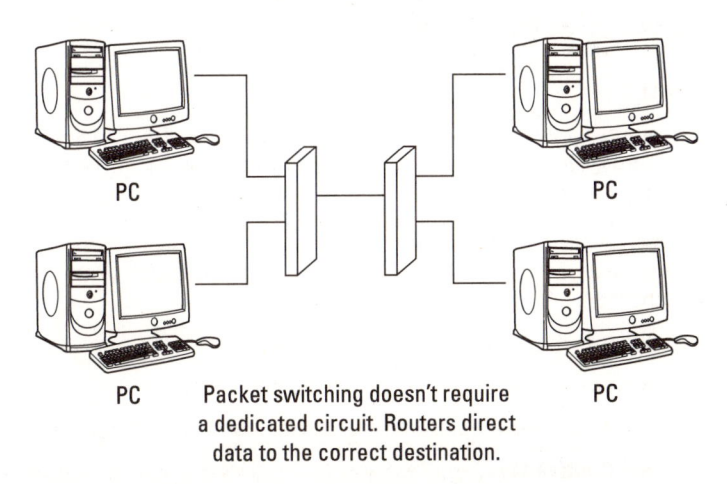

Figure 2-5: Packet switching versus circuit switching

Because users can share bandwidth, rather than each user requiring its own dedicated circuit, a packet-switched network uses network bandwidth more efficiently. IP packets don't have to arrive at their destination in the correct order. Packets contain sequencing information in their address headers. When they arrive at their destination, the receiving host reassembles the packets into the original message.

As packets move through the network, each successive layer of the TCP/IP model adds its own header information. The network layers *encapsulate* the data within these successive headers (see Figure 2-6). The headers contain networking instructions and information, including *from* and *to* addresses. Encapsulation is analogous to placing the data in an envelope and sending it to its destination. Depending on the destination address and intended program, your computer may encapsulate data within multiple envelopes.

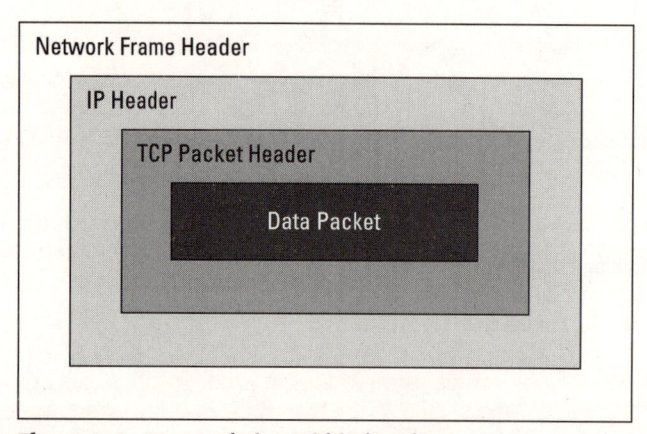

Figure 2-6: Encapsulation within headers

As the encapsulated packet travels to its destination, network layers and services only read the specific header information that instructs them what to do with the packet. As a packet arrives at its destination, each of the headers is stripped away until the data arrives at the destination program. *Demultiplexing* is the process of encapsulation in reverse.

Understanding Network Addresses and Names

For data to reach its destination, it has to have a destination address. The IP describes addressing on TCP/IP networks. The *domain name system* (DNS) translates numeric IP addresses into human-readable addresses. This section describes IP and domain addresses.

Explaining IP addresses

Now it's time for your crash course in Internet Protocol addresses. Don't worry. It is a brief and mostly painless introduction to what IP addresses are and how they function.

IP addresses are a 32-bit dotted decimal, with each byte separated by a decimal point. That is, every IP address is written as four numbers (each up to three digits) separated by a decimal point or dot; for example, 255.255.255.0. Your IP address can be permanent, or your Internet service provider (ISP) can dynamically reassign it each time you reboot or reconnect (using DHCP).

Because of the way data travels over the Internet, each computer must have a unique IP address. For example, your computer's IP address may look something like this:

```
209.179.50.123
```

Demystifying address classes

In the beginning, the powers that be (or in this case, were) divided IP addresses into hierarchal classes. The idea of address classes came before we all had networked PCs; it hasn't worked that well, with class A and B addresses running low.

Class A address numbers range from 1 through 127, class B ranges from 128 through 191, and class C ranges from 192 through 223. (Don't worry about remembering this; it's just nice to know. There won't be a test later.)

- Class A addresses 0.0.0.0 to 127.255.255.255
- Class B addresses 128.0.0.0 to 191.255.255.255
- Class C addresses 192.0.0.0 to 223.255.255.255

The following address ranges are reserved for intranet (internal) addresses that are used on internal networks like your WLAN. These address ranges aren't routable over the Internet.

- 10.0.0.0 to 10.255.255.255
- 172.16.0.0 to 172.31.255.255
- 192.168.0.0 to 192.168.255.255

Exposing restricted addresses

Certain IP addresses are restricted; you can't use them when assigning addresses to your network hosts. These addresses are:

- 255.255.255.255
- 0.0.0.0
- 127.0.0.0

The first number, 255.255.255.255 represents all 1's in binary, and the second number 0.0.0.0 represents all zeros (I know, you guessed that). The first number is a broadcast address. Hosts on a network can use a broadcast address to send data to every host on that network segment. For example, if my subnet, or section of the network that I am connected to is 192.168.0.0, the broadcast address for my subnet would be 192.168.255.255. The second number, all zeros, can't be used because zeros are used to designate a network segment, like my example subnet above (192.168.0.0). I explain subnets in detail in the section that follows.

The third number, 127.0.0.0, is a loopback address, which is used when a machine needs to communicate with itself. No, really, this is used for testing and analysis of network connections. Some machines default to the more common 127.0.0.1 as the loopback address.

Explaining subnets

Administrators often divide large networks into smaller segments called *subnets* or *subnetworks*. Subnets improve performance and security by limiting the number of nodes on each network segment. Subnets are usually traversed by routers (see Figure 2-7).

Each host's IP address consists of two parts: the network prefix that identifies the subnet and the host portion of the address of an individual node. Figure 2-7 illustrates two subnets. In each subnet, the first three numbers separated by decimals represent the network prefix that identifies that subnet. The last number identifies the individual host, which is demonstrated in Table 2-2.

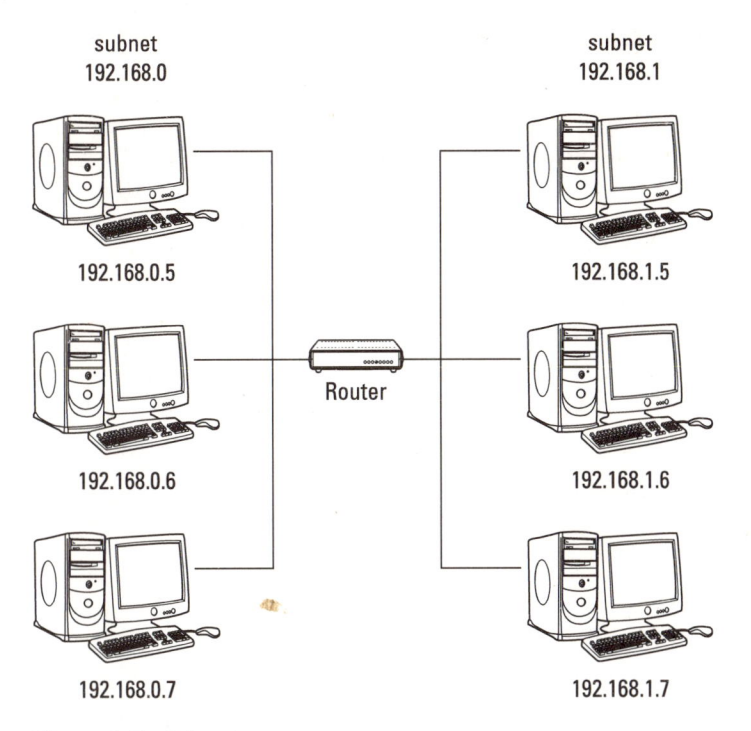

Figure 2-7: Subnets

Table 2-2 Subnets	
Subnet	**Hosts**
192.168.0.x	192.168.0.5
192.168.0.6	
192.168.0.7	
...etc	
192.168.1.x	192.168.1.5
192.168.1.6	
192.168.1.7	
...etc	

Masking subnets

Subnet masking is a technique used in the IP to determine which addresses are available for hosts. The subnet mask is a binary pattern stored on hosts or routers. Your subnet mask represents the portion of the IP address range that refers to subnet addresses, leaving the remaining digits available for hosts' addresses (see Figure 2-8).

255.255.255.0   subnet mask

192.168.122.35   IP address

xxx.xxx.xxx.35   The subnet mask "masks" the first 3 bytes leaving the last byte as the host address.

first 3 bytes used for subnet addresses

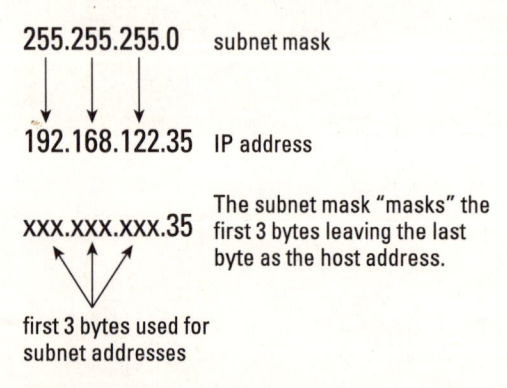

Figure 2-8: Subnet mask

The preceding figure illustrates the default subnet mask (255.255.255.0). The first three numbers (255.255.255) represent the part of the IP address range that identifies subnets, and the last number (0) identifies hosts on that subnet. This means that with the example IP address — 192.168.0.15 — the numbers 192.168.0 identify the subnet. The number 15 identifies the host on that subnet.

Dynamic host configuration protocol

To access the Internet or connect to a *local area network* (LAN), each client must have an IP address. The *Dynamic Host Configuration Protocol* (DHCP) automates this and assigns IP addresses to WLAN clients as they connect to the network. The dynamic addresses change each time a client connects to the WLAN after a reboot, for example, or each client can be "leased" an IP address for a period of time determined by the network administrator (see Figure 2-9).

With DHCP enabled, you don't have to assign an IP address to each machine manually. The DHCP server assigns a random address within a specified range (for example, a subnet). An ISP uses DHCP to assign dial-up users an IP address when they dial in. After the user disconnects, the DHCP recycles the address and assigns it to another user.

198.168.0.65   198.168.0.22

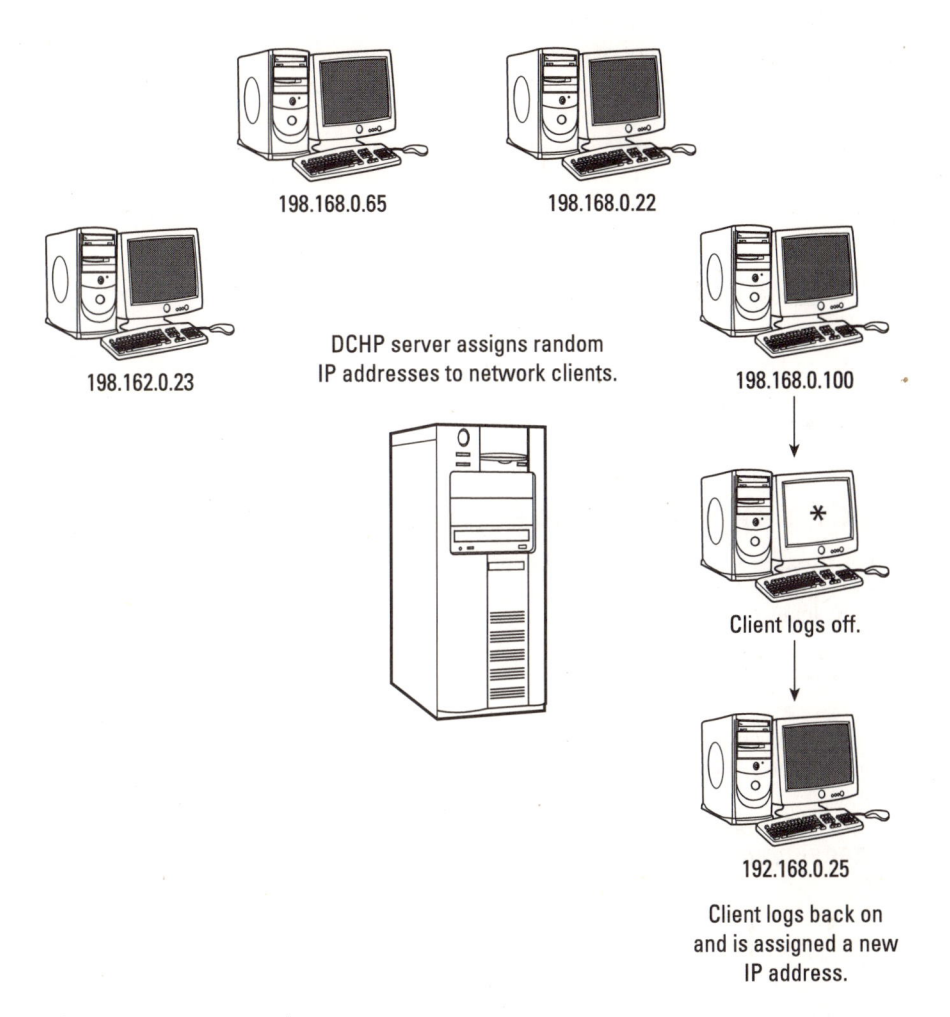

198.162.0.23

DCHP server assigns random
IP addresses to network clients.

198.168.0.100

Client logs off.

192.168.0.25

Client logs back on
and is assigned a new
IP address.

Figure 2-9: DHCP services

Domain names

IP addresses are great for routers and other computers; they work with numbers without any problem. Human beings, on the other hand, have an easier time remembering words than dotted decimal addresses. That's where DNS comes to our rescue.

DNS links IP addresses to human readable domain names. For example, if you wanted to purchase this book (or my other book) from the publisher, it's much easier for you to remember the domain name *wiley.com* than to have to remember the IP address of 208.215.179.146.

DNS servers maintain lists of domain names and their corresponding IP addresses. When you use your browser to visit a Web site, your ISP sends that request through its DNS server, which then looks up the appropriate IP address so that you can communicate with the correct server.

Machines work with numbers, people remember words, and DNS keeps both happy.

Root domains

Domain names are comprised of *labels* separated by dots. The right-most label that follows the last dot is the *root* or *top-level domain (TLD)*. Originally there were seven TLDs; .com, .net, .org, .edu, .gov, .int, and .mil. In 2001 and 2002, seven new TLDs were introduced; .biz, .info, .name, .pro, .aero, .coop, and .museum.

Subdomains and host names

Web addresses include subdomains and host names, as well as TLDs. Working from the far right of the domain name, the network identification gets increasingly specific. The label immediately to the left of the TLD is the domain, which can be preceded by a subdomain (or subnet), and that preceded by an individual host name (see Figure 2-10).

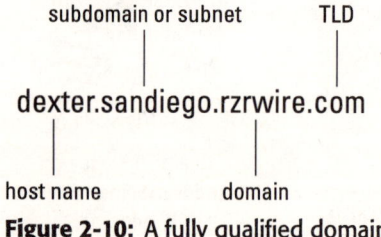

```
       subdomain or subnet      TLD
              |                  |
      dexter.sandiego.rzrwire.com
         |              |
     host name       domain
```

Figure 2-10: A fully qualified domain name including the host

Connecting Networks

It's easier for an administrator to deal with many small networks (or subnets) rather than a single huge one. These smaller segments have to be connected somehow, and bridges, routers, and gateways do just that. On a small network, you probably won't need to use subnets, but you still need to understand bridging, routing, and gateways to connect your network to the Internet and share your internet connection with all the computers on your home network.

Network bridges

A *bridge* (also called a switch) is a device that connects two same-protocol networks, such as Ethernet, and transmits data between them. Bridges have an Ethernet port on each connected network. They are "smart" devices and process each data packet before sending it on to the next network. This prevents errors and corrupt data packets from traveling between subnets.

Until recently there was no method for true wireless-to-wireless bridging on a Wi-Fi network. The new *Wireless Distribution System* (WDS) standard makes this possible. WDS allows traffic to flow from one access point to another as if it were traveling between Ethernet ports on a wired bridge. An access point can still service wireless clients at the same time it's acting as a bridge (see Figure 2-11).

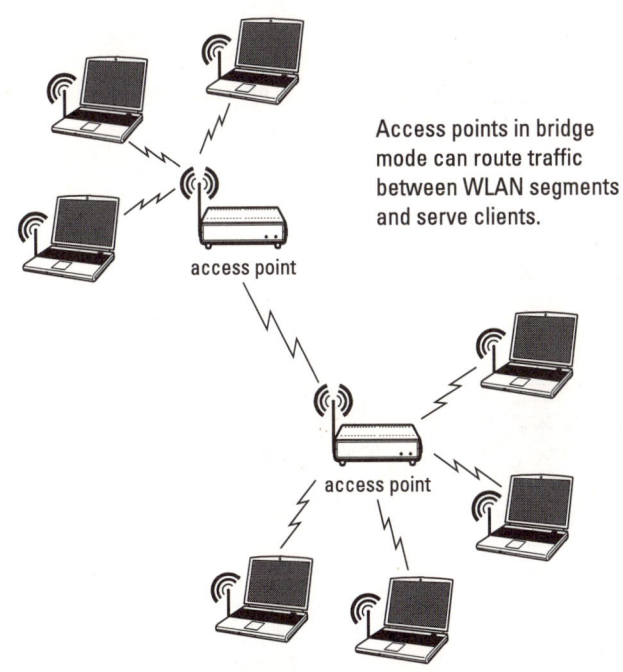

Figure 2-11: Wireless bridging using WDS

Routers and gateways

While a bridge can connect two networks that are alike, routers and gateways can connect different types of networks. *Routers* are hosts that are connected to at least two different networks and can forward data between them. Routers can only route information that's transmitted using a routable protocol such as IP or IPX (*Internetwork Packet Exchange*).

Professionals sometimes use the terms gateway and router interchangeably, indicating a device connected to two or more networks that direct network traffic among them. I draw a distinction between the two. A *gateway* is a device that converts data packets between protocols. A gateway can convert a TCP/IP packet into an IPX packet, allowing computers on the two different networks to "talk" to one another. A router directs traffic from one physical network to another, from Ethernet to Wi-Fi, for example, but does not translate between protocols.

Routers can perform a function known as *network address translation* (NAT). NAT allows you to share one public IP address (assigned by your ISP) among several computers on your WLAN. This allows each host on your network to access the Internet by sharing a single public address. NAT technology conserves Internet addresses and affords some protection to your network by acting as a sort of firewall. Only the IP address assigned to your router is visible to people on the Internet; the rest of your network's internal addresses remain hidden behind the single IP address they are sharing (see Figure 2-12).

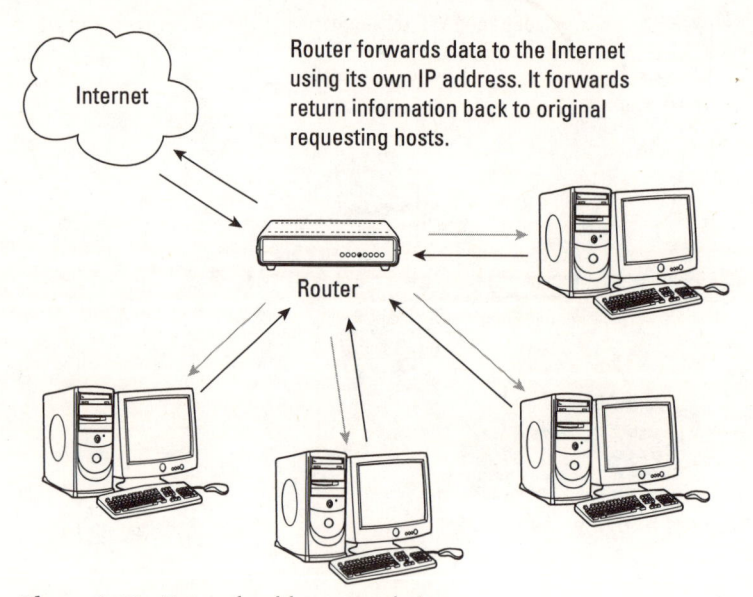

Figure 2-12: Network address translation

To provide NAT services, a router receives each data packet from hosts on your network, and before sending it on to the Internet, changes the header information to reflect its own IP address rather than the originating host's address. The router keeps track of the session information, and when it receives a reply, it reverses the process and sends the data to the originating host.

Summary

In order to understand the threats to your wireless network, you need a basic understanding of networking. In this chapter, I introduced key networking concepts including:

- The basic concepts of networking
- TCP/IP protocol suite and its role in networking
- Network addressing and name resolution
- Connecting networks with bridges, routers, and gateways

Armed with a general knowledge of TCP/IP and networking, you can proceed to an examination of threats to your Wi-Fi network.

✦ ✦ ✦

CHAPTER 3

The People Behind the Problem

In This Chapter

Separating hackers from crackers

The underground cracker culture

What makes crackers tick

Hacker gatherings and conventions

Understanding crackers' tricks

Hacking and cracking resources

Maybe you're wondering why you need to acquaint yourself with the hacker and cracker subculture. Here's why. Acquiring an understanding of these individuals and how they operate will help you separate the hype and hysteria from the real threats to your wireless network.

The media have used sensationalism and misinformation in their representation of hackers. The old and much-used cliché "if it bleeds, it leads" sums it up. Sensationalism often results in high ratings even if it is at the expense of the facts.

The abilities of hackers or the realities of computer threats are exaggerated, contributing to misinformation. The resulting confusion means that the public lacks the facts they need to protect their computers and fewer people take appropriate action, leaving them vulnerable. Frustrated by the lack of accurate information, many people simply give up trying to secure their computers, assuming that they aren't a likely target. Examining who hackers are, the differences between hackers and crackers, and the reality of the threat is the first step toward solving this problem.

Separating Hackers from Crackers

The term *hacker* was originally used to describe a person who was technically proficient with whatever systems he or she hacked and could write exceptional computer code. Some of the original hackers were members of the Massachusetts Institute of Technology (MIT) Tech Model Railroading Club

(TMRC). In fact, the term hacker has been around longer than the personal computer or the Internet, for that matter.

At MIT, students traditionally use the word *hack* to describe the elaborate and often amazing pranks that they play on each other and the faculty. In their world, a hacker is someone who creates something ingenious and truly original.

> **On The Web**
> MIT students maintain a gallery of some of the most ingenious hacks. To view it, pay a visit to: `http://hacks.mit.edu`.

As an activity, hacking has its roots in the computer culture of the 1950s and 1960s. The timesharing of mainframe computers was tightly controlled, and many individuals who wanted to increase their knowledge and technical abilities found alternative ways to gain entry. Out of these roots has grown a full spectrum of hackers from the true law-abiding security experts to computer criminals (crackers).

At one time, the term hacker was a complimentary title bestowed upon someone by peers in recognition of technical abilities and problem-solving skills. This is no longer the case, mostly due to the media's misrepresentation of hackers and crackers adopting the title. The information security community distinguishes between hackers and crackers, labeling crackers as criminals who attempt to exploit those flaws to break into systems.

> **Note**
> Throughout the rest of this book, the term cracker will refer to computer criminals or people unethically breaking into computer systems.

The general term hacker can be broken down into several categories that help define the type of individuals and the threats that they present. These terms are an attempt to categorize human motives and behaviors; but they are not absolutes. They are helpful because they provide a way to discuss different types of individuals and understand their motivations.

Meeting the white hats

Borrowing a simple visual metaphor from old Hollywood westerns, ethical, law-abiding hackers refer to themselves as *white hats*, distinguishing themselves from the bad guys, or *black hats*. Many white hats are security professionals who seek to improve network security. Understanding the software that crackers use and how they operate allows a white hat to take steps to secure a network against likely attacks. Some white hats even create software to defend against known attacks from crackers.

Avoiding the black hats

Using the same Old West metaphor, the computer world's outlaws wear the black hats. A black hat, or cracker, breaks into systems for a number of reasons, with personal gain and bragging rights at the top of the list.

Because the majority of crackers rely on software tools, or *canned exploits*, created by more skilled people, security professionals are able to catch quite a few of them. Unfortunately, when they do catch crackers they often exaggerate the cracker's skill with a computer. This increases misinformation, confusion, and fear among computer users.

While many crackers are inept, it's still important to acknowledge that some crackers are a serious threat to computer security. The technically proficient in their ranks can cause a lot of damage, and the software they create enables unskilled criminals to crack into systems. In addition, many systems are woefully insecure to begin with, and others, such as wireless, have security issues that users are often unaware of.

Understanding the gray hats

Apart from Hollywood's portrayal of the Old West, human behavior is never as black and white as it appears. Considering this, a *gray hat* is an otherwise ethical hacker who, in the interest of improving security, may cross the line occasionally and actually break into a system. While breaking into a system in order to improve security, a gray hat that means well may inadvertently cause damage.

Many gray hats work to find security holes and report them, and they may publicize flaws for bragging rights. Gray hats try to improve security by using publicity to force companies to fix software flaws, rather than giving them a chance to fix the problem before telling the world about it.

Identifying script kiddies

The most common type of cracker lacks any real technical ability or understanding of how computer systems work, and many of them rely on software and scripts created by crackers who are more skilled. Real hackers have many names for these individuals; *script kiddies*, *packet monkeys*, *s'kiddiots*, *lamers*, *warez d00dz* (dudes), and *wannabes*.

In August 2003, 18-year-old high school student and script kiddy Jeffrey Lee Parson made headlines. The FBI arrested Parson, who went by the handle t33kid (teekid), for creating a variant of the Blaster worm. Parson created his version of the worm by editing the *source code* (code used to create the worm) of the original. Because he didn't understand what that code actually did, his version of the worm, dubbed Blaster.B., did little damage in comparison to the original worm.

Further illustrating the problem of sensationalism when reporting computer crime, the press and prosecutors held Parson up as an example of an evil computer genius; even his mother dismissed the idea he was a computer genius. Script kiddies make great headlines, so this pattern tends to repeat itself. They are more of a nuisance than a threat, and it's easy to secure your systems against individuals of their skill level.

Web page defacement, or *cyber vandalism*, is a common pastime among script kiddies. Script kiddies break into a Web server and replace Web pages with defaced versions that they've created (see Figure 3-1). This isn't as hard as you might think, because many Web server administrators don't adequately secure their systems. Web page defacement doesn't take a great amount of skill, just the right tools.

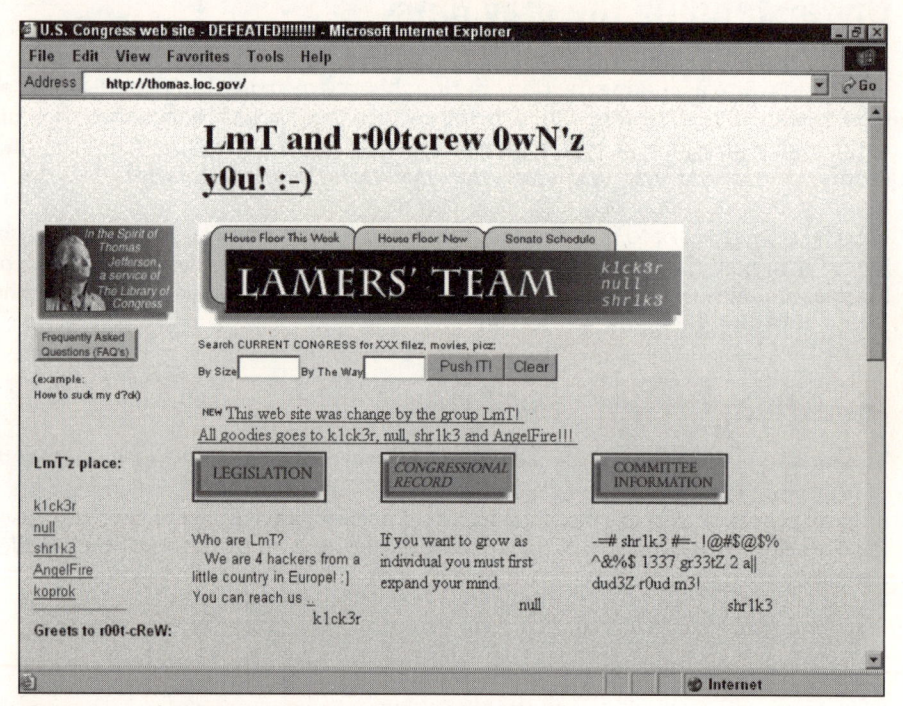

Figure 3-1: The defaced Library of Congress Web site

The average script kiddy in the United States is an adolescent white male, usually intelligent. He likely lives in a parents' basement and collects comic books and *Star Wars* memorabilia (okay, I added that last part). Script kiddies do like to hang out on Internet Relay Chat (IRC) where they organize and brag about their exploits. They also share software on IRC and Usenet newsgroups. The IRC aliases they use can give you a good idea of whom you're dealing with. Most secure, mature adults don't routinely refer to themselves as L0rd Death, Terminator, or CyberG0d.

Based on recent demographics compiled by the FBI and leading security firms, the profile of a skilled cracker has shifted to a professional white male in his early 30s who works with computers (or even security). As attacks become more complex and the stakes get higher, the demographic is shifting toward more skilled, highly intelligent, and well-supported individuals. The threat from cyber terrorists and organized crime is increasing, but the odds are that your WLAN isn't going to be the target of a rogue nation or mafia don anytime soon.

Hacktivists

A *hacktivist* is a hacker-activist who uses cracking as a means to bring attention to a political agenda or social cause. The most common attention-seeking behavior is usually Web page defacement, and the most likely targets are governments and organizations with controversial practices or policies. Web page defacement doesn't require a great deal of skill and many hacktivists are just script kiddies with an agenda.

In 2000, during a hactivism spree, a United Kingdom hacktivist called Herbless hacked the HSBC Bank and various British government Web sites. He did it to protest fuel prices and the government's stance on smoking. His defacement of the Web pages included an activist statement, and on one site he left the following message for the administrator:

```
Note to the administrator: You should really enforce stronger pass-
words. I cracked 75% of your NT accounts in 16 seconds on my SMP
Linux box. Please note the only thing changed on this server is
your index page, which has been backed up. Nothing else has been
altered.
```

Hacktivists on opposite sides of a political argument frequently engage in cyber wars. Indians and Pakistanis routinely hack Web pages from each other's countries, usually referencing the conflict over Kashmir.

Israeli and Palestinian sympathizers have used hacking as a weapon of war. During October-November 2000 this hacking escalated from political to criminal to terrorist. The attack began with the defacement and disabling of more than 30 sites.

Palestinian-affiliated hackers then publicly posted the personal information of the American Israeli Public Affairs Committee members.

Israeli supporters retaliated by posting Palestinian leaders' cell phone numbers, information about accessing the telephone and fax systems of the Palestinian Authority, 15 Internet relay channels (IRC), and an IRC server through which the Palestinian movement communicates. Palestinian hacktivists also attacked several U.S. companies, including the Israeli Public Affairs Committee and Lucent, which has business interests in Israel.

While hacktivism may seem harmless when compared to online credit card fraud and other cyber crimes, it can cause considerable damage to the reputation of a company or agency.

Visiting the Underground Cracker Culture

No chapter about crackers and their cohorts would be complete without a tour of their gatherings and Web sites, and a sample of their language and culture. Think of it as a sort of exchange program, only better, because you don't actually have to meet any of them.

Understanding the background and pop culture of cracking will aid you in protecting yourself, especially against cracker con games such as *social engineering* (mentioned later in this chapter). You might also find it entertaining.

Crackers, particularly script kiddies, have their own language called l33t (*leet*, short for elite). L33t has nothing to do with real hackers, or perhaps more precisely, real hackers have nothing to do with l33t; it evolved among the users of the old BBS systems and later IRC and Usenet. L33t includes many of the following:

- Intentional misspelling of words, including *fone* (phone) and *phreak* (freak)
- Substituting *z* for *s,* usually to denote something illegal, such as in *passwordz, gamez, and warez*
- Using random characters for emphasis: *Whatz up d00dz!#!$#$*
- Obsessive abbreviation: *I got lotsa appz w/docs and codz!#$$#*
- Compulsive shouting with all caps: SO IT L00KS LIKE UR YELLIN!!!

L33t also includes some common letter/number substitutions:

- *4* substituted for *A: as in H4cker*
- *3* substituted for *E,* as in *l33t*

- *ph* substituted for *F*, as in *phreak*
- *1* or *|* substituted for *I* or *L*

> **On The Web**
> For more examples of l33t, visit Chapter 9 of *The Jargon File* (version 4.4.7), by Eric S. Raymond, online at `www.catb.org/~esr/jargon/html/index.html`.

You've probably noticed that l33t has its own unique rules for spelling and grammar. Here I've created an example message in l33t followed with a rough translation:

```
l33td00d: A code monkey wedge his st00pid's gonkulator and the
st00pid is MAD!#!#$$ Lamer can't rel0ad wind0$e - st00pitude
```

Rough l33t to English translation:

```
A low-level programmer broke his boss's expensive and pretentious
new computer and the boss is mad. He doesn't know how to reload
Microsoft Windows on his computer.
```

> **On The Web**
> Visit the L33t-to-English translator at `www.eskimo.com/~mvargas/hax0r.htm`.

What Makes Hackers and Crackers Tick

It's hard to pin down just a few of the things that might motivate a cracker because there are as many reasons for cracking as there are individuals. Still, in interviews and interrogations, a few common motivators continually appear.

> **Note**
> This section focuses on motivations for individuals and doesn't examine political agendas, hate, or other things that might motivate organizations or nations (including terrorists).

Looking for knowledge

Many crackers (and hackers) love information. They tend to hoard and devour it, storing it away for the day when a random bit of technical trivia will enable them to solve some puzzle. In general, if interested in a subject, a cracker will learn everything he can about it. Crackers may break into a system just to learn everything that they can about it.

Fulfilling greed

Financial gain motivates some crackers. Credit card fraud and illegal wire transfers are ways to make easy money. In August 1994, Vladimir Levin, a Russian computer programmer, transferred millions of dollars out of Citibank accounts. He claimed his salary from St. Petersburg's Technological Institute was so low that he had to steal the money to survive.

It's hard to beat money as a motivator. The lure of easy cash, potentially millions, sometimes inspires otherwise ethical individuals to attempt to defraud their own companies. Using computers in financial crime, often committed by insiders, is a major concern for law enforcement organizations around the globe.

Inflating their egos

A major motivation among crackers is status. The bigger the target and the more sophisticated the attack, the more status the cracker gains. To become a l33t hax0r (elite hacker) requires crackers to reach a certain level of knowledge and then demonstrate it to their peers. The amount of media attention that results also provides acknowledgment and increases their status.

The Web sites of government agencies and large corporations are common targets. These targets gain a great deal of attention and status in some cracking groups. In 1996, crackers defaced the Central Intelligence Agency (CIA) Web page, changing the title to Central Stupidity Agency and modified an Air Force Web page to include pornographic images. In October 2000, a young cracker attacked and defaced more than 10 government sites including the White Sands Missile Range, Hanford Nuclear Reservation, and the Department of Veterans Affairs. The list of government pages defaced or disabled is long and distinguished.

Pursuing revenge

Personal revenge can motivate current or former employees to become crackers. This is a growing problem for organizations; some polls claim that approximately two-thirds of network security breaches come from inside the company. These attacks can range from embarrassing to devastating. The employee-turned-cracker may inadvertently crash a system or intentionally destroy information.

Eastman Kodak charged Chung-Yuh Soong, a former employee, with transmitting highly confidential software files to a competitor in California. The only reason the company detected the alleged theft was that the document was so large it crashed the server. At Pixar Animation Studios, the entire company received an e-mail listing the salary of every employee. The e-mail seemed to originate from CEO Steve Jobs' address. Although he did not send it, evidence does point to a current or former employee-turned-cracker.

Hacker Conventions and Gatherings

Hackers, crackers, business people, and even federal agents attend the biggest hacking (or cyberpunk) conventions. Some of these conventions started out as offline meetings of online cracking groups or informal gatherings of hackers.

If you have ever wanted to play a game of Spot the Fed, learn advanced hacks, or just wanted buy some *kewl* T-shirts, go to a convention. The following subsections list some of the top gatherings for hackers and crackers:

DEFCON

DEFCON started in 1993 and gets bigger every year. DEFCON is probably the biggest underground hacking event in the world. It's held in Las Vegas, and the highlights of the convention are the lectures and contests. Contests include the hacker version of Capture the Flag, Spot the Fed, and hacker Jeopardy.

Famous hackers and crackers attend the convention along with the hundreds of obligatory script kiddies. Despite the large contingent of blue- and green-haired kiddies, DEFCON has become a place where security experts, hackers, and even government agents can meet and discuss current security issues.

For more DEFCON information, visit www.defcon.org.

HOPE

2600 Magazine sponsors HOPE (Hackers on Planet Earth). It's held in New York City every two or three years, and is frequented by many of the same individuals that attend DEFCON every year. *2600* posts information regarding upcoming HOPE conventions on its Web site.

Visit 2600 on the web: www.2600.com.

ToorCon

ToorCon (*Toor* is *root* spelled backward; more l33t-speak) has been held in sunny San Diego every year since 1999. Like DEFCON, ToorCon offers several lectures, and although far smaller than DEFCON, it's becoming a worthwhile event to attend.

> **On The Web**
> For ToorCon information, visit `www.toorcon.org`.

2600 meetings

The readers of *2600 Magazine* meet on the first Friday of every month to discuss, learn, and teach each other about technology. All persons are welcome to attend and meetings are held worldwide. *2600* maintains a list of meetings on its Web site.

> **On The Web**
> Visit *2600* magazine on the web at `www.2600.com`.

Worldwide WarDrive

The Worldwide WarDrive (WWWD), now in its fourth year, is a coordinated wardrive (driving around looking for unsecured Wi-Fi access points) to collect Wi-Fi security statistics. Esentially, a lot of people *wardrive* on the same day and then compile the results and post them on the Web site.

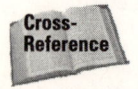

> **Cross-Reference**
> Wardriving is discussed in great detail in Chapter 6.

This site should be of particular interest to you, since wardriving is the principal threat to your WLAN. This Web site illustrates just how many insecure networks there are and how easy it is for wardrivers to find them.

> **On The Web**
> For information about the WWWD, visit `www.worldwidewardrive.org`.

These are just the largest and most well-known conventions and events. Every year, more of them appear, with some surviving and others occurring only once. The DEFCON Web site hosts an updated list of events.

Examining Crackers' Tricks

Crackers have hundreds, maybe even thousands, of tools at their fingertips. Most cracking tools are available online and are free to download and use. The *crack* that a cracker is attempting determines the type of software tool that cracker chooses to use.

Crackers don't rely solely on software on other technological tools. They have many nontechnological techniques at their disposal. These techniques are often even more effective than software tools because companies are less prepared or aware of the threat. This section provides an overview of two effective nontechnical tools, *social engineering* and *trashing*.

Cross-Reference Learn more about technical tools and attacks against Wi-Fi in Chapters 4, 5, and 6.

Social engineering

Social engineering is the art of manipulating people to get passwords or other information. Crackers sometimes refer to it as "hacking the *wetware*" (wetware being the human brain, as opposed to computer hardware). P.T. Barnum once said, "There's a sucker born every minute," and unfortunately, he was right. Now many of those suckers have user accounts, and crackers know it. Sometimes people will give up the incredibly sensitive information to a sincere sounding stranger on the other end of a phone.

Cracker Kevin Mitnik was (and probably still is) an extremely skilled social engineer. Much of the cracking attributed to him involved a telephone, not a computer. People from Motorola, Nokia, AT&T, and Sun Microsystems gave him passwords, phone numbers, voice mailboxes, and even faxed him technical manuals and proprietary information.

A common social engineering trick is to pose as a member of the IT staff. In large companies, it is often easy to pose as staff because there are so many employees that no one knows everyone. To pose as a staff member it helps to have an actual employee name to lend some credibility. A cracker can get the names and extensions of employees from corporate phone books or lists. These also list the names of all other personnel, their departments, and their phone numbers.

A cracker can call Jim in accounting and claim to be Peter from the IT help desk. He can tell Jim that a problem with his system is affecting the whole network. Of course, Jim is now worried and wants to help. Peter asks what password Jim has been using so that he can see if that is the problem. Jim, being the concerned employee he is, gives up his password. The cracker then tells him that this doesn't appear to the problem, and it must not be Jim's system causing the trouble. Jim is relieved and goes about his business. The cracker now has access to Jim's account and starts working from the inside of the network to escalate his privileges to administrator level.

Many popular social engineering tricks involve e-mail. The e-mail, which appears to be from the site administrator, instructs the recipient to run an attached test program. The program then prompts the user to type his password. After the user types his password, the program e-mails it to the hacker. The following is a sample of this type of message:

```
OmniCore is experimenting with an online-high-resolution graphics
display on Windows XP. But, we need you're help in testing our new
product, Turbo-Tetris. So, if you are not too busy, please try out
the Tetris game attached to this email.

Because of the graphics handling and screen re-initialization, you
will be prompted to log on to XP again. Please do so, and use your
real password. Thanks you for your support. You'll be hearing from
us soon!

OmniCore
```

Social engineering is an incredibly effective way of gathering information. The creativity of the cracker and the security awareness of the potential victim are its only limits. There is story after story of crackers gaining vital information with just a simple phone call or e-mail. While social engineering takes a smooth tongue and sharp mind, another technique, trashing, requires less glamorous skills.

Trashing

Trashing is the practice of going through trash to find information. This information can include account names, passwords, credit card numbers, and other security information. Although it is a risky, clandestine process, it can provide valuable information. It's important to remember that what you put in the trash may not stay there. What you may consider innocuous documents or discarded packaging can give an edge to a cracker or a social engineer.

Discarded mail, manuals, or packaging from hardware can facilitate a number of attacks. Crackers can recover sensitive personal information from discarded mail. Manuals and packaging that you have thrown away might indicate exactly what type of Wi-Fi gear you have, assisting a cracker in attacking it.

You might also write down passwords, account numbers, or alarm codes because they are difficult to remember. If you throw these notes away, they can end up in the hands of a cracker. Figure 3-2 illustrates the loot a cracker can find in a trashcan.

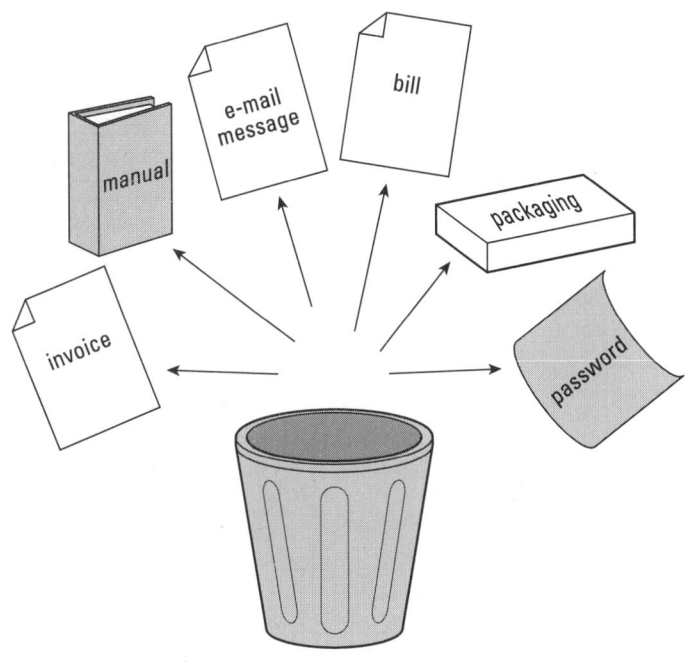

Figure 3-2: Trashing

Entertaining Hacking/Cracking Resources

If you are interested in learning more about crackers and the cracker subculture, you can find many resources on the Internet. On some of these sites, you can download the software tools used by crackers and hackers, but be aware that there is a high likelihood that commercial software offered for download is probably an illegal pirated copy.

I've only included a few sites for you to sample. If you want to see more, do a search online for the words *hax0r* (hacker) or *war3z* (warez) and you'll find hundreds of sites run by script kiddies. If you decide to visit these sites, however, be aware that many of them contain vulgar language, nudity, porn, and adult themes. Others are just bizarre collections of poorly implemented JavaScript effects stolen from other sites.

- **The Cult of the Dead Cow** (www.cultdeadcow.com) — La vaca es muerta... The cow is dead. The cDc has been around since 1984 and in some circles is one of the more respected groups of crackers. Some cDc members have created popular cracking tools, including *Trojans* (malicious applications that masquerade as normal software).

- **Chaos Computer Club** (www.ccc.de) — The CCC is a German organization that began in 1981 and became a formal club in 1986. The CCC Web site is informative and worth a look, but most of it is in German.
- **Attrition.org** (www.attrition.org) — This is a computer security Web site that collects and disseminates information about cracking and security.
- **DoC** (www.dis.org) — Dis Org Crew is a talented and eclectic group of hackers in northern California. Peter Shipley, a prominent member, is the person credited with creating wardriving and coining the term (see Chapter 6).
- **Phone Losers of America (PLA)** (www.phonelosers.org) — The PLA started as an *e-zine* (electronic magazine) devoted to phreaking and computer cracking. The PLA site hosts projects, archives, and a lot of stuff dedicated to crank phone calls.
- **Security Focus** (www.securityfocus.com) — This is one of the most comprehensive security sites that you will find on the Web. While not a hacking site per se, it hosts a detailed vulnerability database that you can reference to help secure your own systems.
- **2600: The Hacker Quarterly** (www.2600.com) — *2600* was named for an early phone hacking exploit (2600 Hz was the frequency used to unlock some phones). Emmanuel Goldstein first published *2600* in 1984. It contains articles about computer cracking and exploiting telecommunications systems. *2600* is interesting, informative, and has been at the center of more than one free speech legal battle.

Summary

In this chapter, I introduced crackers and related computer criminals. The topics I presented include:

- The differences between hackers and crackers
- An introduction to underground cracker pop culture
- The motivations of both hackers and crackers
- A directory of popular hacking conventions and gatherings
- An introduction to some common cracker tricks
- A list of hacking and cracking resources

Understanding crackers, their abilities, and some of their tricks will allow you to make sound decisions to secure your network, and discern between facts and sensationalism regarding computer threats.

✦ ✦ ✦

Hijacking Wi-Fi

CHAPTER 4

In This Chapter

Wi-Fi session hijacking

Spoofing explained

Avoiding rogue access points

Explaining denial of service (DoS) attacks

Crackers have a number of attack methods at their disposal when they decide to go after a WLAN. Many of these attacks require no special technical skill or knowledge; even script-kiddies can perform them. Understanding these attack methods will help you make informed security choices and determine the level of risk (and protection) for your network.

This chapter explains some of the more common and more effective attacks on Wi-Fi networks, but by no means provides a complete list of potential threats. The concepts behind many of these attacks aren't new; some of them have been around a long time and they've been adapted to fit new technology. They apply to wired networks as well as wireless.

Hijacking Sessions

Session hijacking is one class of effective attacks on networks; it isn't difficult, and if the target access point isn't using WEP (or WPA) encryption, then it's extremely simple. Hijacking occurs in many forms on both wired and wireless networks. By hijacking a session, a cracker gains access to a WLAN, where he masquerades as a legitimate user. Session hijacking can also be the first step in other more complicated cracking techniques.

> **Cross-Reference** You can read more about WEP and WPA encryption in Chapter 10

This type of attack targets the *user session* on the WLAN. A user session (or simply session) begins when you connect to and authenticate with an access point (see Figure 4-1) and ends when you log off or the connection *times out*. Connections automatically time out or expire, requiring clients to authenticate again and begin another session. A session time-out can occur after a period of inactivity or after a predetermined amount of time has passed.

Access point authenticates client, and after authentication data is verified and acknowledged, the user's session begins

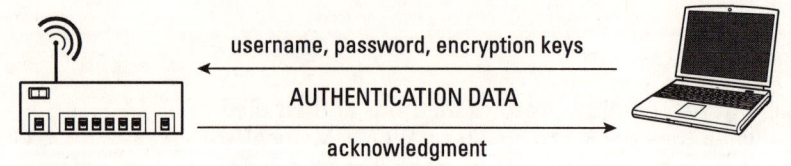

username, password, encryption keys
AUTHENTICATION DATA
acknowledgment

Figure 4-1: Authenticating with an access point

Timing out conserves resources. If sessions didn't time out automatically they would remain open after users leave their computers without logging off. Eventually, a server might be overwhelmed tracking thousands of open sessions for clients that aren't actually there. It also has a security application; closing timed-out sessions helps prevent unauthorized persons from hijacking sessions that have been left open by authorized users.

> **Note** When you connect to a WLAN, or to your ISP for that matter, your session periodically times out (typically in an hour), and your computer authenticates again without your intervention. The process is largely invisible to the user.
>
> Similarly, when you log on to a Web site, you begin a session that runs for a specified amount of time or until you log out. If you leave a Web site without logging out, you may return a few minutes later to find out that you are still connected. This happens because your session hasn't yet expired.

Spoofing

Contrary to popular belief, spoofing is not a method of communicating anonymously on the Internet. Generally, using spoofing alone, you can only communicate one way (sending data). The cracker uses a *sniffer*, which is a software application that sniffs or passively listens to network traffic. The attacker waits for someone to authenticate with the access point and captures the authentication data (sequence and acknowledgment numbers, see Figure 4-2).

The cracker can then insert commands into the data stream, spoofing a legitimate user's IP address so that it appears that the inserted packets originated from that user's machine. The cracker inserts commands that force the target server (or access point) to reestablish the connection and then hijacks the session by authenticating with the sequence and acknowledgment numbers that he sniffed (see Figure 4-3).

Chapter 4 ✦ **Hijacking Wi-Fi** 61

client

access point

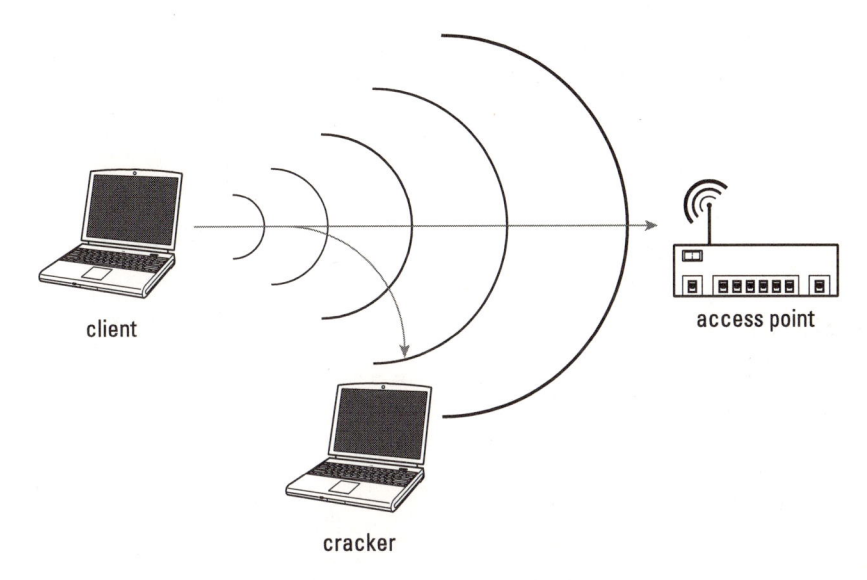

cracker

Cracker uses wireless sniffer application
to listen to and intercept all traffic on a WLAN

Figure 4-2: Sniffing network traffic

> **Note** Spoofing allows a cracker to send data to network hosts with which he normally couldn't communicate. All of which are a component of session hijacking, a *man-in-the-middle* attack, and *denial of service* (DoS) attacks.

An attacker can emulate the access point and send a legitimate client a *disassociate frame*. A *frame* is a *packet* of data in network communications. Engineers and programmers use the two terms interchangeably. The disassociate frame disconnects the client from the WLAN. When this happens, the attacker can spoof the client's MAC address and take over the user's session. The session remains open because the access point didn't send the disassociation message, the attacker did. As far as the access point is concerned, the original user is still connected and authenticated (see Figure 4-4).

> **Cross-Reference** Read how crackers can use disassociation messages in a type of denial of service (DoS) attack in Chapter 5.

This type of attack exploits a *race condition*. In this case, the attacker forces the legitimate user to disconnect and then *races* to take over the user's session. If the attacker can spoof the victim's MAC before the client authenticates again, he can hijack the session and take over until the session times out.

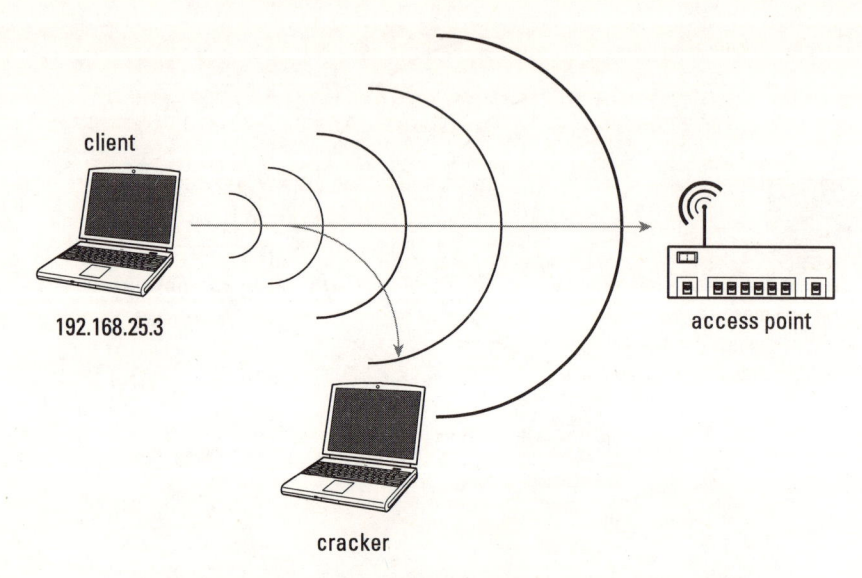

First a cracker sniffs the network to identify a valid IP address and intercept that user's authentication data

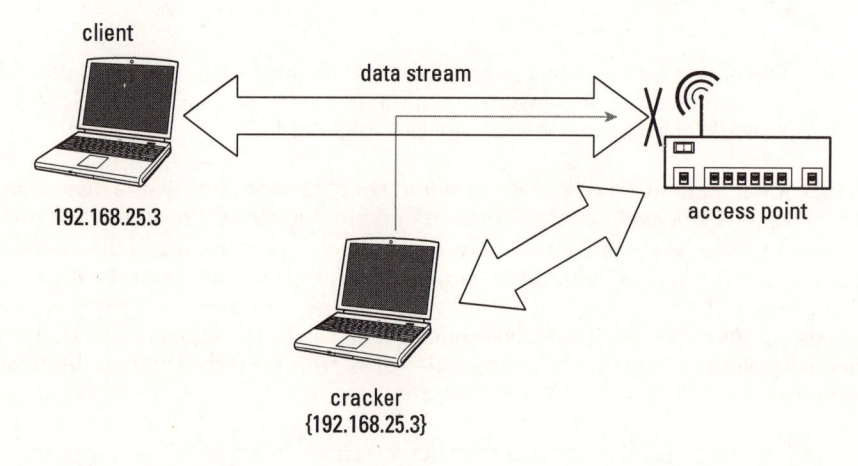

Using the legitimate client's IP address the cracker then inserts data into that user's session, tricking the access point into reauthenticating the user. The cracker quickly authenticates with the stolen authentication data and assumes the legitimate client's identity.

Figure 4-3: Session hijacking

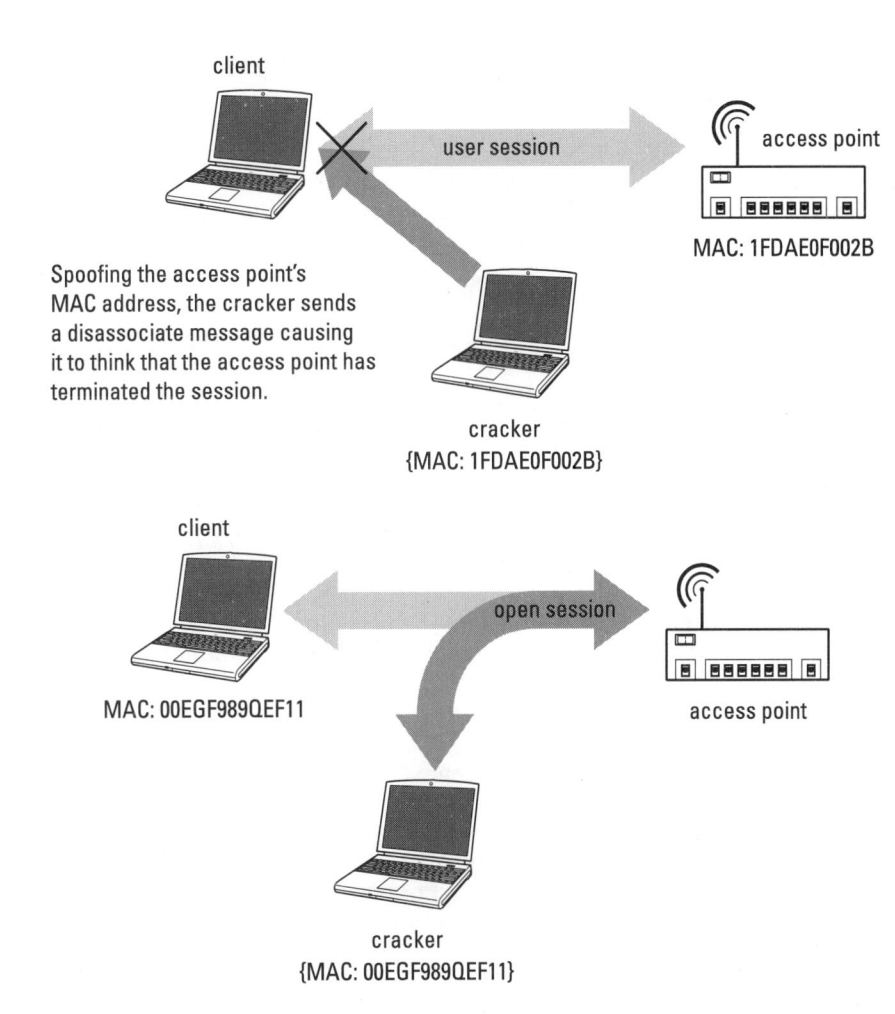

Spoofing the access point's MAC address, the cracker sends a disassociate message causing it to think that the access point has terminated the session.

Before the client reassociates, causing the previous session to close, the cracker masquerades as the client using the client's MAC address and assumes control of the previous session.

Figure 4-4: Spoofing a MAC address and exploiting a race condition

Tip: Use WEP or WPA encryption on your WLAN to prevent this type of session hijacking and most forms of IP spoofing from occurring.

Caution: Routers and firewalls can also prevent many types of spoofing attacks, but due to flaws in the 802.11 protocol, crackers can use spoofing when attacking WLANS.

Explaining race conditions

A *race condition* occurs when a device, application, or system attempts to perform multiple functions at the same time, but must perform them in a specific order for them to be successful. The system depends on the timing of events, and if the timing is off, unexpected results can occur.

In the example I used in the previous section, a race condition occurs when a cracker forces a user to dissociate with an access point. The legitimate user attempts to reassociate, and at the same time, the cracker is trying to spoof the MAC address of that user and connect with the access point. The outcome depends on the timing of the two events. If the cracker manages to associate before the legitimate user reestablishes a connection, he can take control of the session. If the legitimate user reestablishes a connection first, the cracker is out of luck.

Race conditions are not exclusively security concerns; they occur in many different areas of information technology. Race conditions occur in software when different functions attempt to read or modify the same data simultaneously. This can cause the computer or applications to crash or lead to other unexpected results.

A cracker can sometimes exploit this type of race condition and get one function to overwrite or modify the data used by another function, the result being that the cracker can gain access to a system or application. Race conditions aren't easy to exploit, and often a cracker has to attempt this kind of attack multiple times before it is successful.

Public hotspots

While it's possible that session hijacking could occur on your own WLAN, public Wi-Fi hotspots are a more likely environment for these (and other) attacks to occur. Many free public or *open* hotspots don't use WEP or WPA encryption, which would prevent most session hijacking attacks. (Chances are you're using some form of encryption on your WLAN. If not, I hope that you will after reading this book.) The administrators of these WLANs want users to be able to connect freely and with little trouble.

I'm specifically talking about public hotspots, meaning they are free and open for the public to use, not the commercial hotspots hosted at major chain restaurants, coffee houses, and other gathering places. Commercial hotspots generally employ some form of encryption (beyond WEP and WPA) and are less likely targets for many session hijacking techniques.

However, beyond session hijacking, there are many other techniques that a cracker can use to target clients. Hotspots, both public and commercial, present an opportunity for a cracker because there are so many, often unsecured, Wi-Fi clients present. A cracker can target data stored on clients as well as data broadcast between clients and access points.

A cracker can configure his computer to masquerade as a legitimate access point to intercept and record data sent to that access point. WEP may deter casual eavesdropping and spoofing, but if the user authenticates with the cracker's access point, WEP is no protection here. This amounts to a form of wireless sniffing, and the cracker can collect all data sent through the user's machine, including usernames, passwords, and credit card numbers.

> **Caution** When connecting to an access point at a hotspot, make sure the access point is legitimate. See the section "Understanding Rogue Access Points" later in this chapter.

While using a hotspot, unsecured client machines are also susceptible to malicious software, including worms, and viruses, that can spread via a network. Crackers can also use *blended attacks* when targeting other network clients. Blended attacks use malicious software to target known security vulnerabilities. An example would be a worm that targets a security flaw in an operating system in order to gain access to a computer and then delivers a *payload* such as a virus or installs a program that allows the cracker to access and control the machine.

> **Cross-Reference** To read more about viruses and protecting yourself from them, see Chapter 7.

Protecting yourself at hotspots

There are a number of things that you can do to protect yourself when you're using a public or commercial hotspot. These include:

- Using a *virtual private network* (VPN)
- Installing a personal firewall on your notebook or PDA
- Installing antivirus software
- Using encryption when sending files across the network

Using a VPN will help keep all of your data private when you're working with e-mail or sensitive data. A VPN uses encryption to create a *tunnel*, which is a secure connection between two computers, even over the public Internet (see Figure 4-5). In fact, using a VPN allows business people to connect securely to their company's e-mail and servers via the Internet. VPNs are usually available only to business users.

A VPN creates a secure (private) connection between two points across the public Internet

encrypted "tunnel" for secure communication

the Internet

corporate server

client

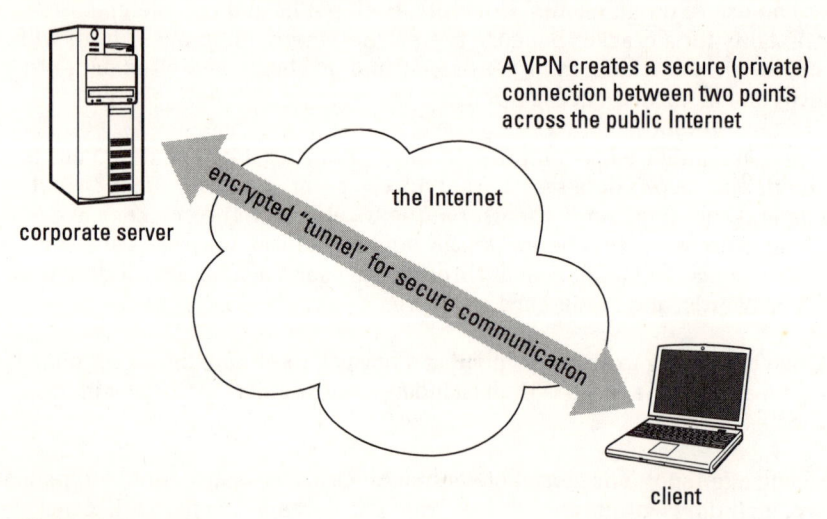

Figure 4-5: A VPN

In order to use a VPN, you or your company needs to have a VPN solution installed at your end of the connection, and you need to have VPN client software installed on your machine. Usually, you also have to have a routable public (Internet) IP address, not an IP address reserved for internal use on a WLAN or intranet. This is because many VPN solutions won't operate over NAT and require a public IP address, whether it's static or assigned dynamically by DHCP.

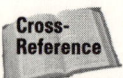

Cross-Reference For more information about NAT and DHCP see Chapter 2.

Once you connect to your VPN and authenticate, the VPN encrypts all of the data between your machine and the server. This secures any communication with your corporate network, such as e-mail and file transfers, but once you connect to the Internet you've left the protection of your VPN behind because you're now communicating with the ISP servers and not your company's VPN (see Figure 4-6).

One exception to this would be if your company supplies you with an HTTP proxy server that you can access over the VPN. A proxy server receives Web page requests from your browser, forwards those requests to the company's ISP, and sends the pages back to your browser (see Figure 4-7).

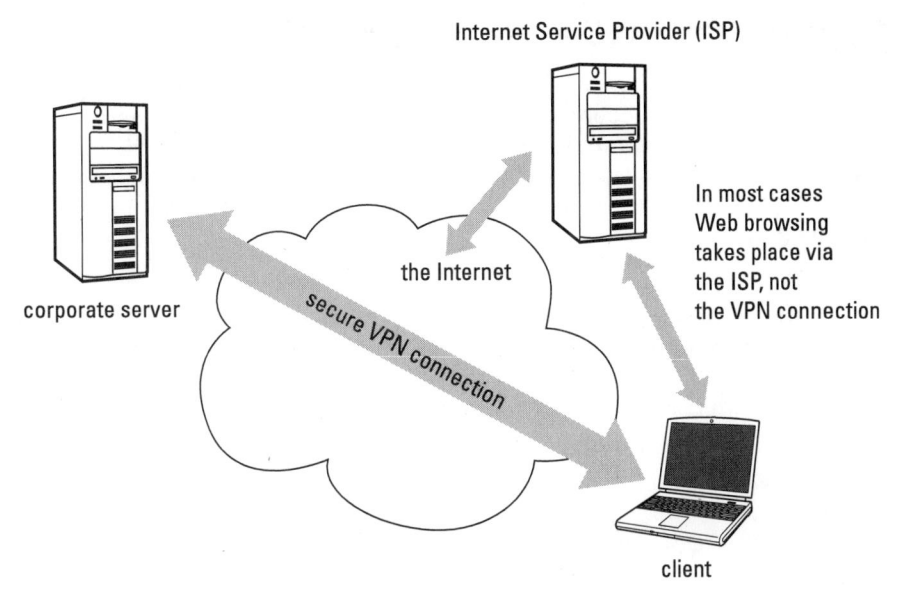

Figure 4-6: A VPN doesn't secure Web browsing

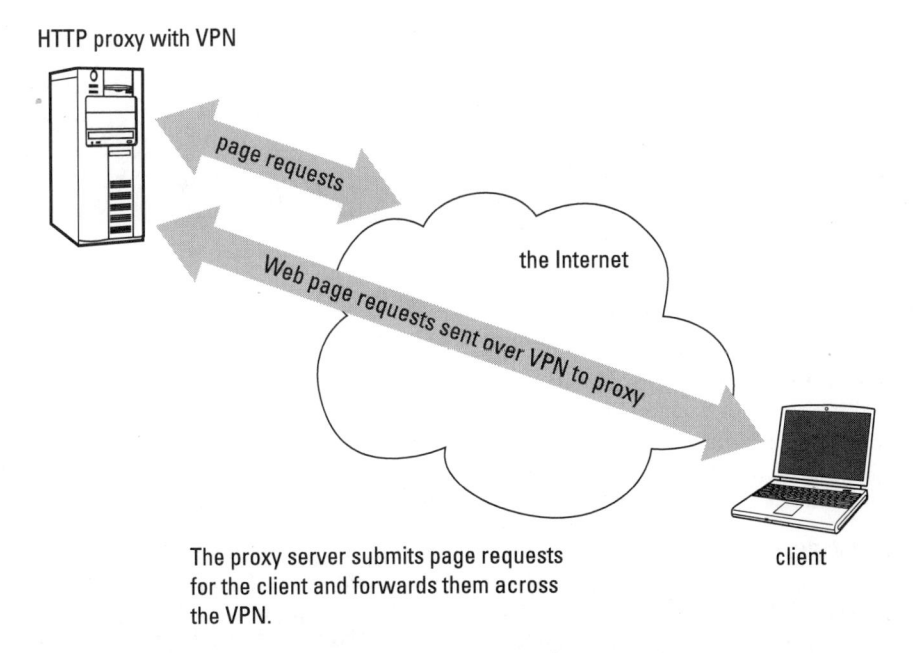

Figure 4-7: An HTTP proxy server

Exposing the man in the middle

As already mentioned, spoofing is a type of *man-in-the-middle* attack. A man-in-the-middle (MITM) attack occurs when a cracker spoofs the MAC or IP address of a network client or access point. Masquerading as a legitimate node on the network, the cracker can intercept or inject data into the communications stream between two other nodes (see Figure 4-8). Usually, the affected nodes will be unaware that this is happening.

In a man-in-the-middle attack, a cracker can intercept data or forge data to make it appear that the message originated with a legitimate client

client
IP: 192.168.25.3
MAC: 1FDAE0F002B

access point

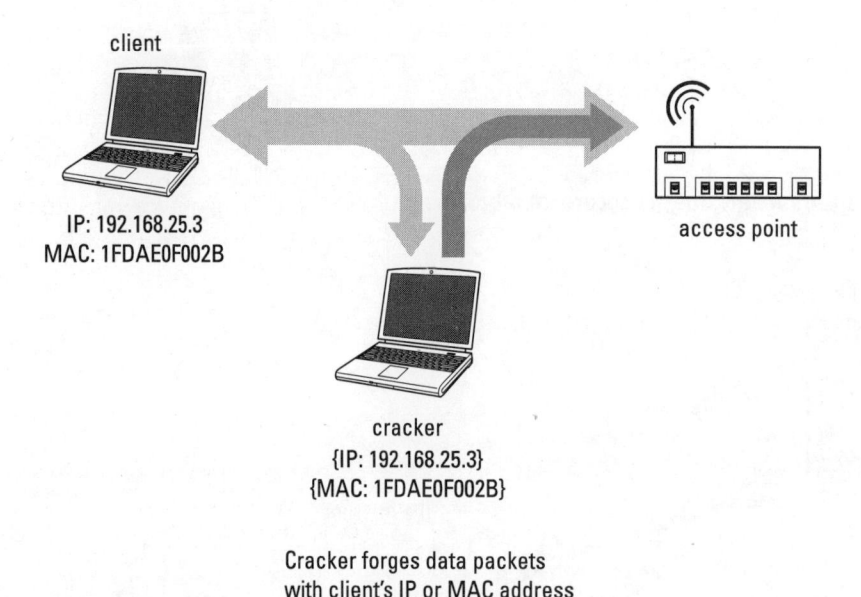

cracker
{IP: 192.168.25.3}
{MAC: 1FDAE0F002B}

Cracker forges data packets
with client's IP or MAC address

Figure 4-8: Man-in-the-middle attack

Understanding Rogue Access Points

A rogue access point can be a couple of different things. Often, you'll hear the term used when referring to an unauthorized access point added to a corporate network by an employee. Usually, the employee is on a wired *local area network* (LAN) and wants wireless connectivity. Perhaps his company hasn't adopted wireless or doesn't plan to, so the employee takes it upon himself to install his own access point at his desk and connect it to the corporate LAN. This is a problem, and it can create serious security issues for the company's IT department.

Another definition of a *rogue access point* is an access point set up by a cracker to collect data such as passwords, usernames, or credit card numbers. This is the one I will concentrate on. It is a growing threat in public places and one that you need to take seriously. A rogue access point masquerades as a legitimate access point, often as part of a commercial wireless network. Incidents involving rogue access points have been popping up all over the country, particularly in urban areas and in places where business travelers congregate.

Wi-Fi hotspots are quickly becoming ubiquitous across the country. Companies like Wayport, Boingo, and T-Mobile offer service nationwide. Crackers take advantage of peoples' implicit trust in these commercial wireless offerings. They do this by setting up rogue access points that look like access points belonging to commercial Wi-Fi networks.

The cracker sets up the rogue access point and names it with the same SSID as a commercial access point. The name might be Wayport or T-Mobile. When you are near one of these access points and you start up your wireless device, you'll see the rogue access point, but because the cracker has used the name of your wireless service provider's legitimate access points, you connect to it (see Figure 4-9).

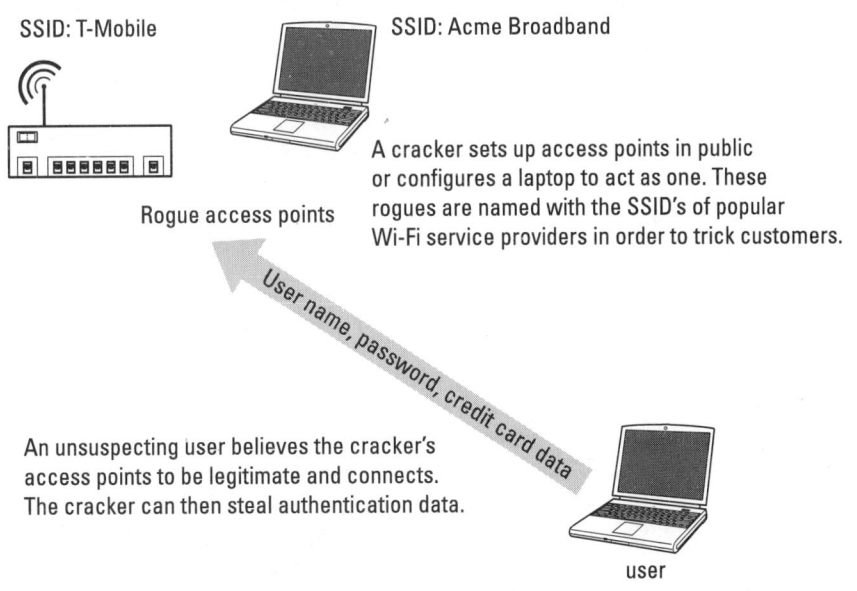

Figure 4-9: A rogue access point

The rogue access point can eavesdrop on all of the traffic sent to it by client machines. WEP or WPA encryption offer no protection here because you are associating with the access point and machines on both ends of the connection have the encryption key and can decrypt all of the data.

The cracker may also configure the rogue access point to provide a new-user registration service so that it can collect credit card numbers and personal information from unsuspecting people who think they are signing up for wireless service on a legitimate provider's network.

Unfortunately, there is little that you can do to expose this type of wireless attack. IT personnel in companies can use hand scanners to detect and locate access points added to their networks. They know if an access point should be there or not, so if they detect one they know it's a rogue.

Unfortunately, there isn't a similar solution for discovering rogue access points masquerading as legitimate ones. A scanner is expensive, and even if you know how to use one, all it does is tell you that an access point is there, not whether or not it's legitimate.

The best method of protection is to be cautious and educate yourself. Download a list of the locations of all hotspots and access points offered by your carrier. Look for signs that indicate your carrier offers service at a particular location. When you turn on your laptop and see an access point available or if there is a sign present that indicates there should be an access point there, check to see if it is on the hotspot list you downloaded. If neither of these conditions is true, be cautious and don't connect to the access point in question. To protect yourself at a hotspot, you should follow the instructions that I gave earlier in this chapter.

Uncovering Denial of Service Attacks

Denial of service (DoS) attacks can flood a device with network traffic or exploit other weaknesses in order to prevent users from connecting to and using a service (see Figure 4-10). DoS attacks can involve a single attacking computer, or a cracker may use many compromised computers (called *zombies*) to simultaneously attack. This simultaneous attack is called a *distributed denial of service attack* or DDoS. By themselves, DoS attacks don't steal information or compromise machines, but crackers use them as a component in many spoofing attacks.

There are many different DoS attacks that target wired and wireless networks, operating systems, and applications. DoS attacks aren't limited to computers; people have launched DoS attacks against the telephone systems of companies or organizations with controversial policies.

A cracker overwhelms an access point with thousands of tasks or a large amount of network traffic, preventing legitimate users from connecting to the network

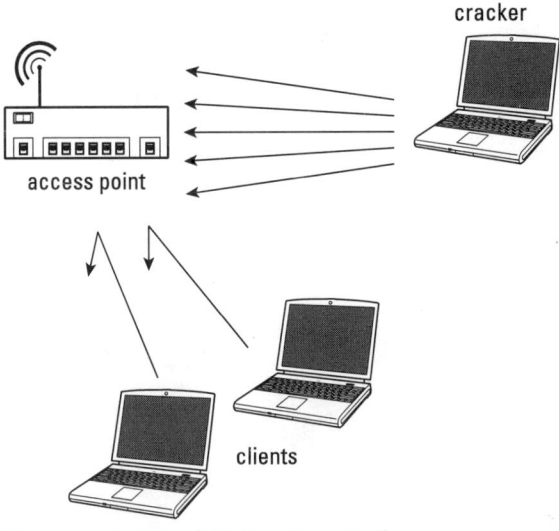

Figure 4-10: A denial of service attack

In one scenario, hundreds of protesters continually dial the toll-free number of the target organization, staying on the line as long as possible when the target answers and immediately redialing if they get a busy signal or are disconnected. This prevents legitimate customers or members of the target organization from calling the number, effectively denying them that service. It also results in a huge phone bill for the target company.

Some DoS attacks aren't even intentional; flaws in networking devices can cause them. In 2003, a flaw in some Netgear routers caused them to continually poll the Internet time servers at the University of Wisconsin, inadvertently causing a DoS attack on those servers. In this case, the flaw in these products coupled with the number of the products in use (hundreds of thousands) created an enormous problem for the university's IT personnel.

> **On The Web**
>
> For a firsthand account of this accidental DoS attack and to see if your Netgear device is one of the flawed models visit www.cs.wisc.edu/~plonka/netgear-sntp/.

I'm going to concentrate on DoS attacks against wireless networks, specifically some of the more likely or common attacks. When used as part of a spoof attack, a cracker may launch a DoS attack against an access point to prevent users from connecting to it while simultaneously spoofing the access point and inviting clients to connect to his machine (see Figure 4-11).

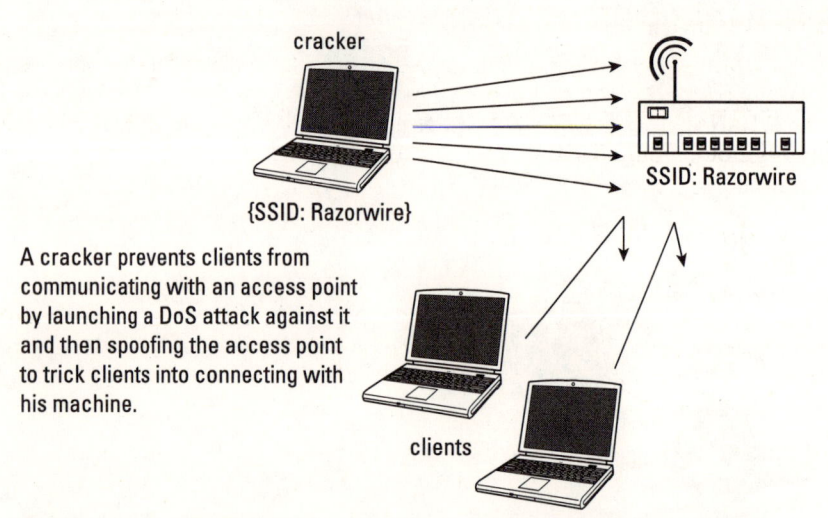

A cracker prevents clients from communicating with an access point by launching a DoS attack against it and then spoofing the access point to trick clients into connecting with his machine.

Figure 4-11: DoS attack against an access point as part of a spoof attack

Unfortunately, unless you are relatively network savvy and experienced in configuring firewalls, these DoS attacks are hard for home users to defend against. Fortunately, it's unlikely that you will find your home WLAN the target of a DoS attack; public hotspots are the more likely victims.

Many of these issues will only be resolved by changes to firmware and addressing problems in the actual 802.11x protocols. To better protect your system, stay aware of updates for your hardware and apply patches immediately when your vendor makes them available.

WPA denial of service

The new Wi-Fi Protected Access (WPA) encryption is a more affective replacement for the flawed WEP algorithm used in many Wi-Fi products. WPA authenticates users logging on to the network to prevent unauthorized persons from connecting. WPA has a feature to prevent active attacks; if it senses because of repeated failed authentication attempts that it's under attack, the access point shuts down and resets. This leaves users without wireless connectivity during this time.

A cracker can take advantage of this feature and turn it into a DoS attack. By sending repeated packets of unauthorized data to the access point, a cracker can trick

WPA into thinking that it's under attack, forcing it to restart. The attacker can do this repeatedly, creating a denial of service.

Disassociate frame attack

In this type of DoS attack, the cracker repeatedly sends spoofed dissociate frames that force the access point and clients to repeatedly disconnect, effectively shutting down the WLAN. This is the same technique discussed in the "Hijacking Sessions" section at the beginning of this chapter, only on a grander scale.

Strong signal jamming

In this simple and direct DoS attack, an attacker uses a wireless transmitter to broadcast interference on the same frequency channel used by the access point, effectively drowning it out and creating so much radio frequency (RF) noise that Wi-Fi devices in the area can't operate (see Figure 4-12).

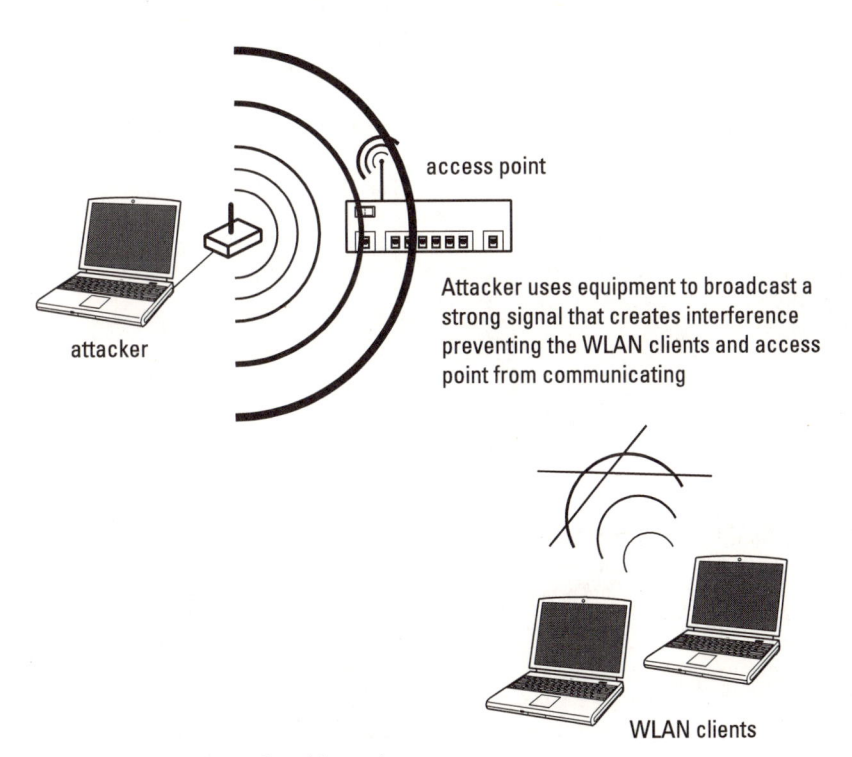

Figure 4-12: Strong signal jamming

Because many devices share the same frequency as Wi-Fi networks, this DoS attack may be mistaken for RF interference, and the WLAN administrator may not even know the network is under attack.

Because of the signal power required, however, the attacker often has to be in somewhat close proximity to the access point and can be located using a portable hand scanner. There really isn't a defense against this attack other than switching radio channels and trying to locate the source of the interfering signal.

FakeAP flood

FakeAP is a software tool designed to defend a WLAN against wardrivers and crackers. FakeAP works by creating hundreds, or even thousands, of virtual access points, hiding your real access point among them. Crackers and wardrivers have no idea which access point is real, so your WLAN is effectively hiding in plain sight.

Unfortunately, FakeAP also works great against hotspots. A new twist on the DoS attack turns this defensive tool into one for attacking. The attacker, with a wireless notebook running FakeAP, locates himself near a hotspot and runs FakeAP to create thousands of fake access points. Visitors to the hotspot don't know which access point is real and can't connect to the WLAN (see Figure 4-13).

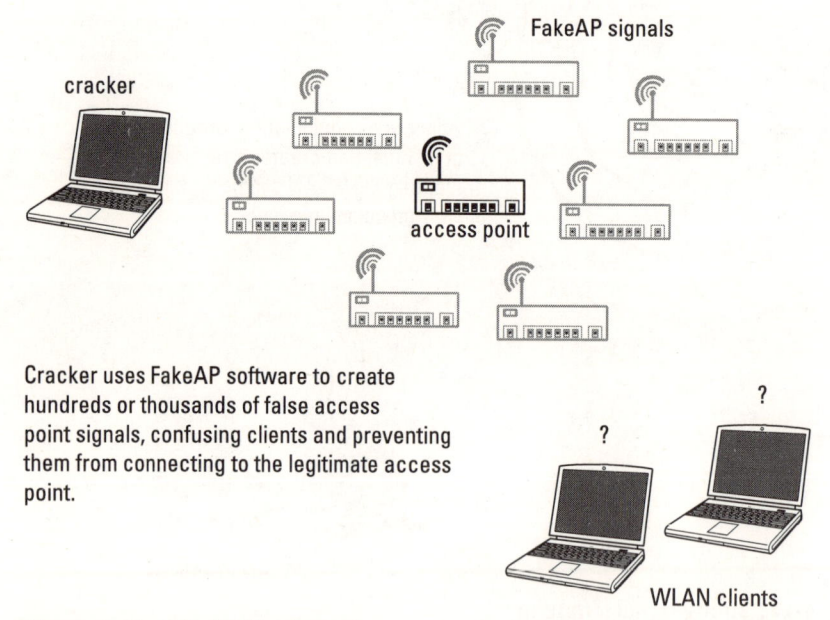

Cracker uses FakeAP software to create hundreds or thousands of false access point signals, confusing clients and preventing them from connecting to the legitimate access point.

Figure 4-13: FakeAP flood attack

Summary

In this chapter I introduced some of the methods that a cracker might use when attacking your WLAN. These included:

- How a cracker might hijack a user's Wi-Fi session
- How crackers use spoofing to gain access to wireless networks
- What rogue access points are, and how crackers use them to steal user logon information
- Denial of service (DoS) attacks, including how crackers use them to disrupt wireless networks

These are just some of the attacks that are possible against wireless networks. In the next chapter, I will present some others, including attacks that target WLAN clients like your notebook or even your PDA.

✦ ✦ ✦

CHAPTER 5

More Wireless Attacks

In This Chapter

Attacks against WLAN hosts

Exploiting OS vulnerabilities

Known security vulnerabilities

Understanding wireless sniffing

In addition to attacks against the insecurities in Wi-Fi, crackers can also attack host computers on your WLAN, often using attack strategies carried over from wired networks. Once crackers have access to your WLAN, they can proceed to attack network hosts directly. Insecurities in the Windows OS don't go away when you unplug the Ethernet cables; they're just easier to get to. This chapter further examines ways that a cracker can compromise hosts and clients on a Wi-Fi network.

Attacking Hosts

After gaining access to your WLAN by attacking the access point, or worse, by simply connecting to it because you've failed to secure it properly, a cracker has free reign and can use your Internet connection, or attack your computers directly.

Many home users fail to secure computers on their home network adequately. They often take steps to protect their networks from attacks that originate from the Internet, by adding a firewall to their Internet connection (see Figure 5-1), but installing a WLAN gives crackers another avenue of access.

By attacking your computers via your WLAN, a cracker is essentially coming in the back door and bypassing the security you've set up to keep Internet crackers out; at this point, the cracker can use many attacks that also work on Ethernet networks. Only the network medium has changed, not the vulnerabilities of the devices that are connected to it.

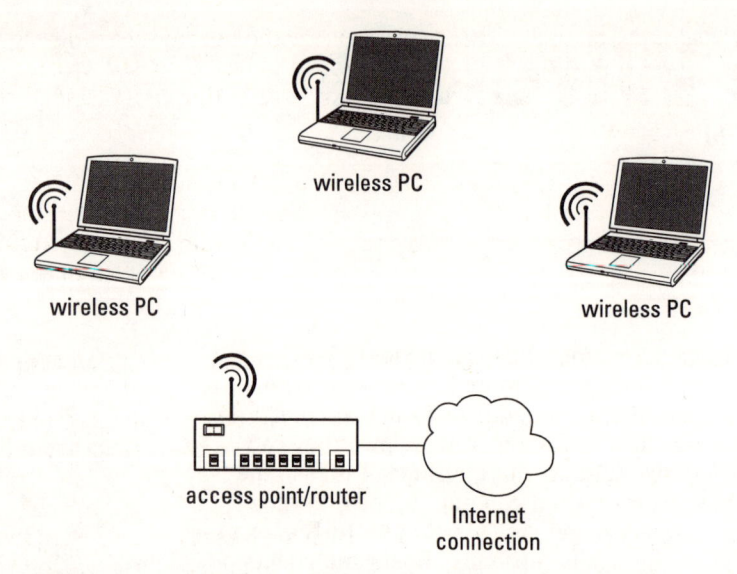

Figure 5-1: Protecting a network from Internet attacks with a firewall

The first step a cracker can take is to use a TCP/IP *port scanner* to identify hosts on your network, and determine which avenues of attack are open. A port scanner is a software application that attempts to contact computers via a range of IP addresses. The cracker knows the IP address range, or subnet range, of your WLAN. The cracker discovered this by identifying default IP settings that you may have failed to change, by connecting to your WLAN via DHCP, or by sniffing your network.

> **Cross-Reference**
> Chapter 11 addresses securing your WLAN, changing default settings, and identifies problems with DHCP. Chapter 4 introduced network sniffing. Chapter 2 discusses IP addresses, and introduces DHCP.

Computers using TCP/IP assign numbers to certain services or *ports*. Computers use these port numbers when connecting to a service on another machine. For example, when you request a Web page via your Internet browser, your browser connects to the Hypertext Transfer Protocol (HTTP) port on the remote Web server, also known as *port 80*.

Common port numbers range from 0 to 1023, major services like *file transfer protocol* (FTP, port 20), and Simple Mail Transfer Protocol (SMTP, port 25) for outgoing mail are common ports. Internet applications can also use ports numbered from 1024 through 49151.

Using the default address range the cracker configures the port scanner to scan these addresses. For example, if your IP address range were 192.168.0.1 thru 192.168.0.254, the cracker would scan all of the addresses in that range to discover computers on your network.

A port scanner sends data packets to each IP address in the range, waits for a response, and records the results. If a computer responds to a port scanner, the cracker knows that there is a host at that address, and depending on the scanning software and technique, can determine the computer's operating system and identify vulnerabilities.

A cracker can use a port scanner to determine if a specific service is running on the target computer. If the cracker wanted to determine if the target was running a Web server, the port scanner would attempt to connect to port 80 and listen for a reply. If there were Web server software running, it would reply and the cracker would then know the application name and version number.

Port scanners are available online, and many of them are free. The most effective way to defeat this cracker reconnaissance technique is to install a personal firewall on your computer. A personal firewall will prevent a port scanner from communicating with ports on your computer. This is often called *stealth mode*. In stealth mode, attempts to port scan your computer go unanswered, making it appear to a port scanner that there is no computer at that IP address.

Cross-Reference: For more information about personal firewalls, please read Chapter 11 in this book.

On The Web: Once you've installed a personal firewall, you can test its effectiveness by visiting www.grc.com and running the ShieldsUP! utility. This will scan the ports on your computer, inform you of the results, and make suggestions for improving your security.

The computer you are testing needs to be connected directly to the Internet. If you are sharing an Internet connection among several computers, the broadband router or connection sharing software you are using will prevent ShieldsUp! from testing your computer.

You can also run ShieldsUp! before you install your personal firewall, just to see how vulnerable you are.

Using OS weaknesses

One avenue that a cracker can use when attacking your computer is to attempt to exploit vulnerabilities in your computer's operating system. Vulnerabilities arise due to software bugs, or design errors that may allow a cracker to crash or take control of a machine.

Every OS has its share of problems, but Microsoft Windows has more known vulnerabilities than most other operating systems. In part, this is due to the large number of computers running Windows operating systems of some sort. Some estimates put this number at over 94 percent of the home computer market.

A large installed user base makes Windows the prime target for crackers and security researchers who expend plenty of effort finding security holes in it. Crackers want more bang-for-their-buck, and discovering and exploiting security holes in Windows gets them plenty of attention. Security researchers recognize this and attempt to find and close holes before crackers do.

A portion of the problem with Windows vulnerabilities is attributable to Microsoft's poor track record with security issues. Until recent years, the company did little to secure their products, and placed most of the emphasis on usability and quick delivery. The result was that many Microsoft products shipped with plenty of existing security problems. However, this is changing and Microsoft is placing greater emphasis on security issues and proactively identifying security holes in its products and addressing them quickly.

Microsoft closes security holes in its products by releasing software patches that users can download and install. These patches update the portion of Windows affected by the security problem, correcting the issue and securing the system.

On The Web To check if your Windows computer is patched, and up to date visit `http://v4.windowsupdate.microsoft.com/en/default.asp`.

In order to make sure that your system is up to date and secure, you should enable Windows Update. Windows Update will automatically check for new patches and alert you when they are available. You can even configure Windows Update to download and install patches automatically.

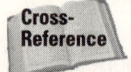

Cross-Reference Chapter 11 details securing Windows XP and provides instructions for activating Windows Update.

Note If you would like to receive timely information regarding security matters that affect Microsoft's products, you can sign up for the Microsoft Security Update e-mail alerts at `www.microsoft.com/security/bulletins/alerts.mspx`. There are two levels of alert that you can choose from. The Microsoft Security Update e-mail alerts are aimed at home users, and the Microsoft Security Notification Service provides more technical information geared toward advanced users and IT professionals.

Another vulnerability related to Windows XP Home Edition is *Simple File Sharing*. Simple File Sharing allows you to share files and folders with other people on your WLAN. Unfortunately, Simple File Sharing provides no access control whatsoever, so anyone connected to your network can access your shared files and folders. If a cracker can connect to your WLAN and browse, then the cracker can access your shares and you would never even know.

Unlike previous versions of Windows, you can't even assign a password to protect a share. Only Windows XP Professional Edition provides access controls for shared folders. However, there is one thing you can do to protect your shares: You must assign a password to the guest account.

Windows XP uses the guest account to manage connections to your shares and your computer. Windows gives you the option of turning off the guest account, but this doesn't solve the problem, because the account remains active. Windows can't completely disable the guest account; it's integral to too many networking functions.

However, if Windows uses the guest account for Simple File Sharing, you should be able to assign a password to that account and by default, protect shared files and folders, as well as shared printers, from unauthorized use.

Unfortunately, Windows doesn't allow users to assign a password to the guest account via the User Accounts control panel (see Figure 5-2). To do this you'll have to enter the password through the command line interface.

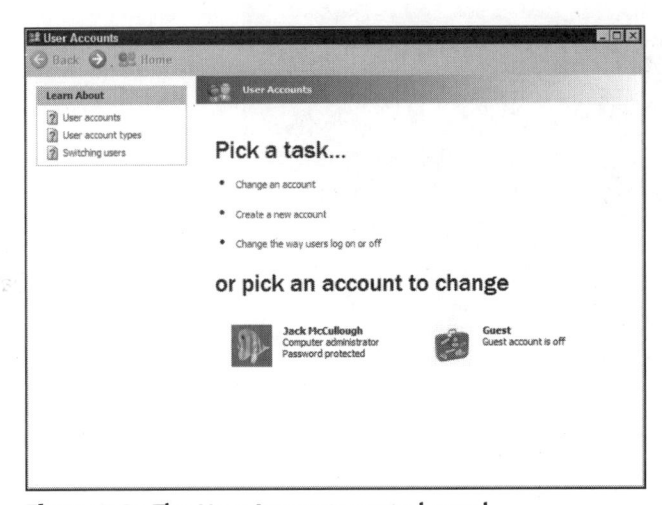

Figure 5-2: The User Accounts control panel

STEPS: Assigning a password to the Windows XP Guest account

1. **Click on the Start button.** The Start Menu appears.
2. **From the Start Menu, choose Run.** A small Run window opens (see Figure 5-3)
3. **In the Run window, type** CMD **and click OK.** A command window appears.

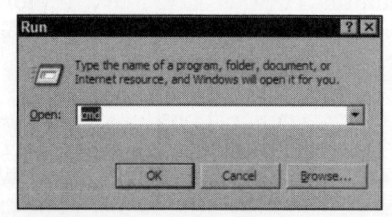

Figure 5-3: The Run dialog box

4. **In the command window, type** net user guest "password" **as shown in Figure 5-4.** Replace "password" with your chosen password, and do not include the quotation marks.
5. **Press the enter key.** Now, whenever anyone attempts to connect to your shared resources and folders they'll be prompted for a password.

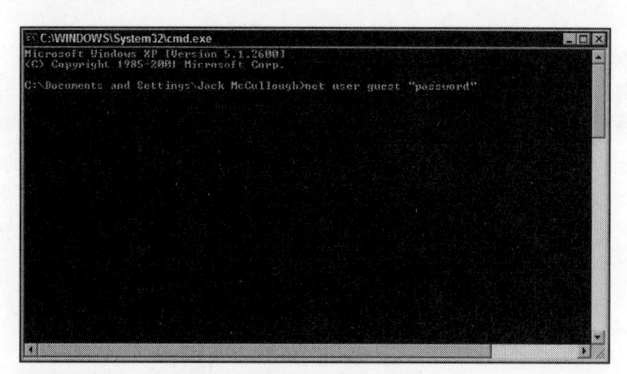

Figure 5-4: The command window

Using known security issues

Another avenue of attack is for the cracker to exploit known security issues in applications and devices. Like OS vulnerabilities, these security issues exist due to design flaws, software bugs, or through unexpected results arising from the integration of different software products.

By some estimates almost 90 percent of computer security incidents result from an exploit of a known security issue. That is, an issue that was publicized and that a vendor corrected via a patch, and yet many people failed to apply the patch and fix their own systems.

Manufacturers, researchers, and crackers regularly discover new vulnerabilities. Manufacturers and software developers release patches to correct these problems when possible, but unlike Windows OS patches, there isn't a simple automated way for a home user to collect all of these updates and install them.

Many software developers and hardware manufacturers offer security or patch bulletins delivered via e-mail. Like Microsoft's security update e-mails, these alerts notify you when a company has identified a security issue and has released a patch or devised a workaround. If a developer offers e-mail updates and bulletins consider subscribing.

Caution No company ever sends its customers patches via e-mail. If you receive an e-mail that directs you to install an attached patch, delete it even if it appears to be from your software's developer.

This is a common ploy that crackers use to get unsuspecting computer users to install malicious software. You can be sure that the patch attached to the e-mail is dangerous software designed to take over, damage, or spy on your computer.

Some software applications have automated update features, and when possible you should use them. This is a common feature in many security products, such as antivirus software. Symantec's Norton Antivirus, has a live update feature that not only downloads new virus information, it also keeps the application current, and will alert when upgrades and patches are available (see Figure 5-5).

Cross-Reference For more information about viruses and antivirus software, please refer to Chapter 7.

If your software doesn't have an automated update feature at all, or if it has this feature but it doesn't support patching the application itself, you'll have to locate and download your patches manually, which can be a time-consuming task requiring a great deal of vigilance to keep your systems secure.

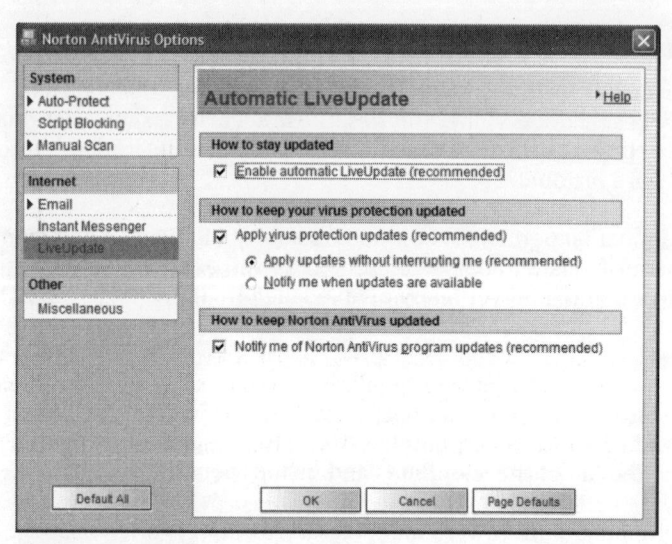

Figure 5-5: Configuring the Norton Antivirus live update feature

Unfortunately, there aren't too many security sites that are aimed at the average home user. However, I can recommend two Web sites that you can use to research patches and vulnerabilities for your applications and devices. These are:

- www.securityfocus.com
- www.cert.org

Security Focus is a respected security site that hosts one of the most comprehensive databases of vulnerabilities available. You can search by device, application, operating system, and vendor. Using this database you can research security issues related to your devices and software.

Security Focus also offers quite a few security-related newsletters and mailing lists. While most of these are geared toward IT professionals, there are a couple of lists that you may find useful, including a Windows-specific security newsletter.

> **On The Web**
> Visit www.securityfocus.com/newsletters if you would like to subscribe to Security Focus newsletters.

The *Computer Emergency Response Team* (CERT) Coordination Center also keeps a database of vulnerabilities and incidents, as well as a host of other useful information. CERT also has some useful security information available for home users.

> **On The Web**
> For more information, visit the CERT Coordination Center at `www.cert.org`.

As you download and install patches, it's helpful to keep a record of your patch history, so that if you ever need to reinstall an application, or you suddenly start having problems you'll have a record of what patches you've installed.

> **On The Web**
> CERT offers a helpful patch checklist available as a PDF download at `www.cert.org/homeusers/HomeComputerSecurity/checklists/checklist2.pdf`.

Check for new patches often, including whenever you purchase or download a new application. It's a good idea to go to the developer's Web site and check that you have the most up-to-date version. A lot can happen between the time the software is produced and the day that you purchase it. It may have been sitting on a shelf for months, and in that time, the developer may have released several patches.

Sniffing the Network

Whether or not you have secured your WLAN, with wireless networks, it's not easy to predict the propagation of radio waves. Radio waves do not travel the same distance in every direction because obstacles such as walls, people, and land formations create *attenuation,* or weakening of the Wi-Fi radio signal. Because of this, many wireless networks are susceptible to attack by wardrivers using sniffer software and equipment.

> **Cross-Reference**
> I discuss placement of WLAN equipment and dealing with interference issues in Chapter 8.

Sniffers are available in many flavors, and most of them are free software downloads for anyone to install and use. People can use sniffers to detect and map out your WLAN, but also to capture data and even crack your encryption.

> **Cross-Reference**
> I discuss weaknesses of Wi-Fi encryption in Chapter 10.

Sniffers are available for most operating systems, including handheld pocket PC devices, and even sniffing for WLANs using a keychain Wi-Fi seeker gadget (see Figure 5-6).

Figure 5-6: You can use devices like the Chrysalis WiFi Seeker to sniff out WLANs.

How sniffing works

Although you have likely heard the argument that nobody owns the airwaves, wireless sniffing is a form of electronic eavesdropping. Essentially, to determine the location of a WLAN and if that WLAN has security precautions in place, an intruder using a Wi-Fi network card, and some free sniffing software can roam until he identifies a signal. By analyzing its strength, he can alter his geographic position until he finds the target access point (see Figure 5-7).

Some of the sniffing software that an intruder uses can also capture passwords, unencrypted data, and even aid in cracking encrypted data. Because wireless communications are accessible to anyone connected to or within range of the WLAN, sniffing is the most effective technique in wireless attacks.

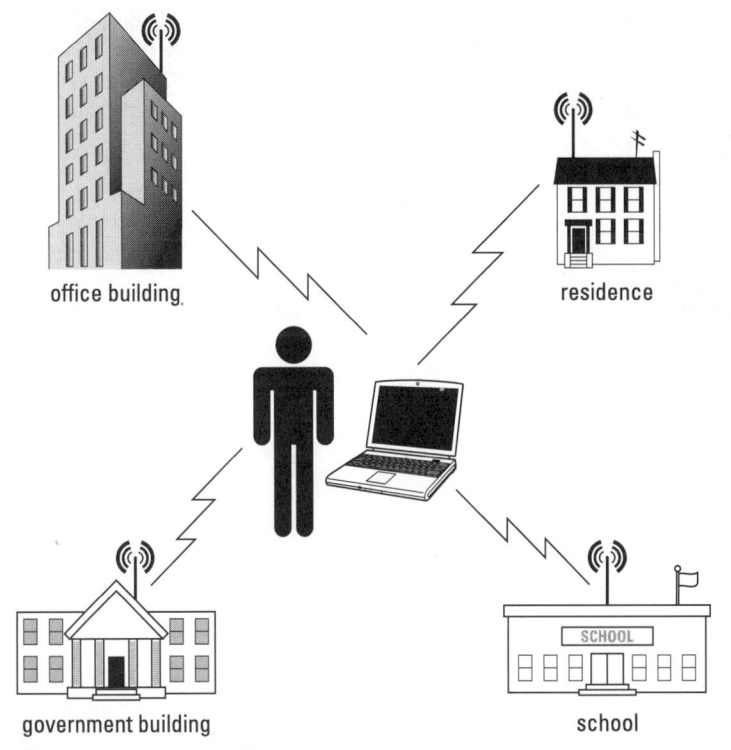

Figure 5-7: Roaming for targets

The promiscuous NIC

Typical sniffers are software programs that passively intercept and copy all traffic that passes by the system's wireless network interface card (NIC). Most sniffers are used for legitimate functions such as network monitoring and troubleshooting. Sniffers are an invaluable tool for diagnosing network problems during communication hosts and nodes. A sniffer captures the data coming in and/or going out of the NIC and displays that information in a table. Some sniffers can even take that a step further by analyzing the data and looking for potential security weaknesses and other network problems.

To do this, sniffers place NICs into a monitoring state called *promiscuous mode*. The term implies that the NIC does not have to participate in the network or send packets and, therefore, can passively eavesdrop and capture data unnoticed. This is like a hidden wiretap that allows third parties to listen in on network transmissions. For this reason, sniffers can wreak havoc on a secure network because they are almost impossible to detect.

> **Note** Wireless sniffers are powerful tools that you can use to audit and defend a WLAN as well as attack it. Because sniffing is usually a passive activity there is little you can do to detect and prevent it. Your best defense is to refer to Chapter 11 and follow the steps for securing your WLAN, and Chapter 10 for best practices regarding Wi-Fi encryption.

Summary

In this chapter, I reviewed some additional avenues of attack that crackers are likely to exploit, and steps that you can take to thwart them. These topics included:

- Attacks against network hosts
- Exploiting OS vulnerabilities and known security issues
- Keeping Windows patched
- The secret to securing Simple File Sharing
- Patching known security vulnerabilities
- How wireless sniffing works

By following the recommendations in this chapter and throughout the book, you'll be better prepared and you will be less likely to find yourself the victim of a wireless attack.

✦ ✦ ✦

CHAPTER 6

Wardriving

Wardriving has become a pastime for Wi-Fi geeks and crackers all over the world. Whether they're looking for a network to crack, free and anonymous Internet access, or to find insecure networks for statistical purposes, you should know how it's done so that you can take steps to protect yourself. This chapter examines the tools and motivations for wardriving.

Introducing Wardriving

Wardriving is the act of driving around with equipment that can be used to detect Wi-Fi access points located in homes and businesses (see Figure 6-1). It doesn't require a large investment in equipment or a high degree of technical skill. Anyone can wardrive, and with the proliferation of Wi-Fi networks in homes and businesses, there are plenty of wireless networks waiting to be found.

In the days of dial-up computer Bulletin Board Systems (BBS), before the Internet or the World Wide Web, a cracker would use a piece of software called a *wardialer* to dial a huge range of numbers, one after the other, and listen for a computer to answer at the other end. The wardialer would then log the number, attempt to connect, and, if successful, identify the computer and service at the other end of the line. Wardriving gets its name from comparisons to this cracker activity.

> **Cross-Reference** To read more about crackers and other computer criminals see Chapter 3

The majority of *wardrivers* are nothing more than curious Wi-Fi geeks; technophiles who love wireless equipment and enjoy driving around, and discovering and then mapping networks. The members of many *Wireless User Groups* (WUGs) wardrive to collect data and compile statistics about wireless use and level of security or insecurity of discovered WLANs.

In This Chapter

Understanding wardriving

What are warchalking and warspying

Software for defending your WLAN

More about wardriving

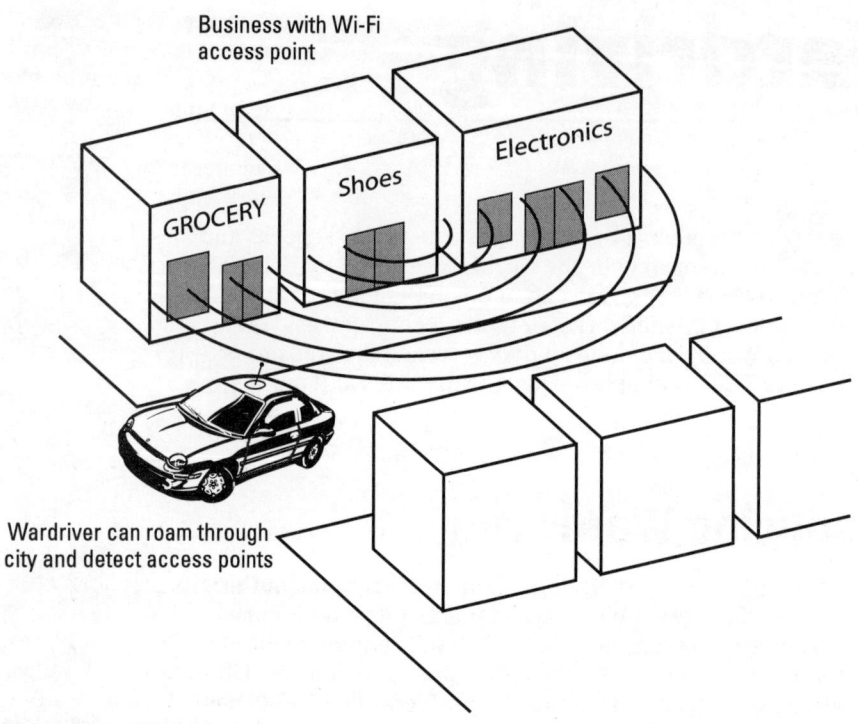

Figure 6-1: Wardriving

However, harmless geeks aren't the only people wardriving, and you need to protect your network from the minority of wardrivers that look for unsecured access points for purposes that are more nefarious.

Crackers wardrive to find vulnerable networks that they can break into. Once crackers get inside a WLAN, they can connect to the Internet or attack other computers without law enforcement or security personnel tracking and apprehending them. Efforts to locate the source of the attack lead back to the WLAN from where it originated by the cracker, but not directly to the cracker. This raises a question of liability. Can the courts hold you responsible for crimes that a wardriver commits while using your WLAN without your knowledge? In most cases, the answer is no. In order for you to be criminally liable for the actions of a wardriver, law enforcement must be able to demonstrate that you had intent or that you arranged for a cracker to access your network in order to facilitate the crime.

> **Note** Laws relating to wardriving vary among states. See the section "Legality of wardriving" later in this chapter for information regarding different state laws.

However, depending on the circumstances and the laws of your state you could be civilly liable for the actions of a hacker. For example, if you have a small business and a cracker was able to steal credit card information because you failed to take reasonable steps to secure your network and protect customer data, you may be subject to a lawsuit by customers who've had their personal information stolen.

Because of the potential for liability arising from misuse of your WLAN or from theft of personal information, it's important that you take steps to secure your WLAN. While you may believe you're unlikely to be the victim of a cracker, given the popularity of wardriving the chances are greater than you may think.

How wardriving works

Wardriving doesn't require a great amount of technical skill or expensive equipment. Many people have much of what they need lying around their house already. All a would-be wardriver needs before hitting the road is:

- A laptop with a Wi-Fi adapter or built-in wireless capability
- Wardriving software (available free on the Internet)
- A high gain antenna (optional)
- A power adapter for running the laptop off the car's battery (optional)

Once the wardriving software is running, all a wardriver has to do is drive around any commercial or residential area and he can detect WLANs. Using a Windows XP laptop and an application named NetStumbler, I was able to detect several WLANs in a rural area of Southern California in just a few minutes (see Figure 6-2). This was without the addition of an external antenna. Had I been in a more populated area I might have discovered hundreds of wireless networks.

After identifying a wireless network, a cracker can attempt to connect to it. Unfortunately for WLAN user, connecting to an unsecured WLAN is trivial even for a technical newbie. If you fail to take steps to protect your network, someone may connect to it and use your Internet connection, or access files on your computers, which is why it is so important to know how to protect your network.

Hiding your network

One of the first steps you can take to help protect your wireless network is to turn off the *service set identifier* (SSID) broadcast on your access point. The SSID is the network name that identifies your WLAN, and access points regularly broadcast data packets, called *beacon frames*, that announce their presence to the world. Clients and wardrivers can listen for these beacon frames and detect the presence of your access point, as well as discovering your SSID.

Many access points allow you to disable the SSID broadcast. This is a good first step, and it will prevent some wardriving software from detecting your WLAN. However, beacon frames aren't the only wireless packets that contain the SSID, and tech-savvy wardrivers have software that listens in passive mode and can collect and analyze other network traffic to recover the SSID and detect your access point.

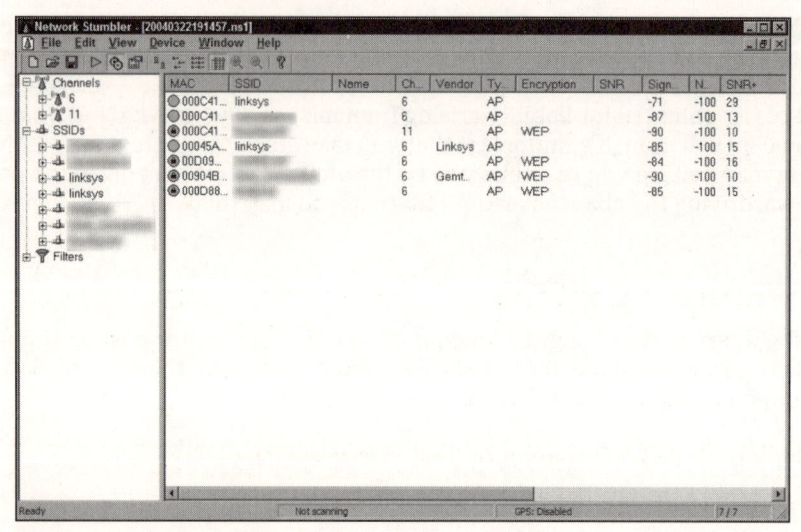

Figure 6-2: NetStumbler detecting wireless networks

Still, disabling your SSID broadcast can help hide your network from many casual wardrivers, and if they can't see it, they can't connect to it.

Changing default settings

Failing to change the default settings on access points is the main reason so many WLANs are at risk. Every piece of networking equipment comes configured with pre-set or default settings. Manufacturers configure their hardware to facilitate use, not security. They want the device to be simple to set up and use right out of the box, which minimizes customer frustration and cuts down on calls to support centers.

Ease of use doesn't equal secure, however; it usually results in the opposite. Although devices may have security features built in, manufacturers often assume that users will be able to activate and configure these after they have the device running. Few people ever take the time to enable security features on their access points, let alone reset the default settings that they probably don't realize are a problem.

A cracker can use the default settings on your access point to identify the make and model of the device, connect to your network, and even take control of it. Default settings for most access points are available online, so you can assume that every cracker knows them. Changing these default settings is an important first step toward protecting your network.

Cross-Reference: The steps for changing default settings and securing your network are discussed in detail in Chapter 11. After you read the information presented here, I suggest that you proceed directly to Chapter 11 and follow the instructions to secure your network.

Using the default SSID

Every access point has a default SSID assigned by the manufacturer. Even if you have disabled SSID broadcasting, you should change the default SSID. If you haven't changed the default SSID, a cracker can use it to identify your access point's manufacturer and model. For example, the default SSID for most access points manufactured by Linksys Inc. is *Linksys*.

A cracker who knows which hardware you are using can break into your network with less difficulty. Don't make it easy for an intruder to gain access. Follow the best practices in Chapter 11, and change your SSID to something unique that doesn't identify you or your network hardware.

Passwords and usernames

Once a cracker knows which access point you are using, he can use the Internet to look up the default user names and passwords for your particular brand of device. Like the default SSID, the default username and password is public knowledge, and many crackers probably know them by heart.

Unless you change the settings, a cracker can connect to your access point, often through a Web browser (see Figure 6-3), and change the device's configuration. Your username and password are like the keys to your network; you wouldn't give out the keys to just anybody, would you? Take the time to change the default username and password following the steps explained in Chapter 11.

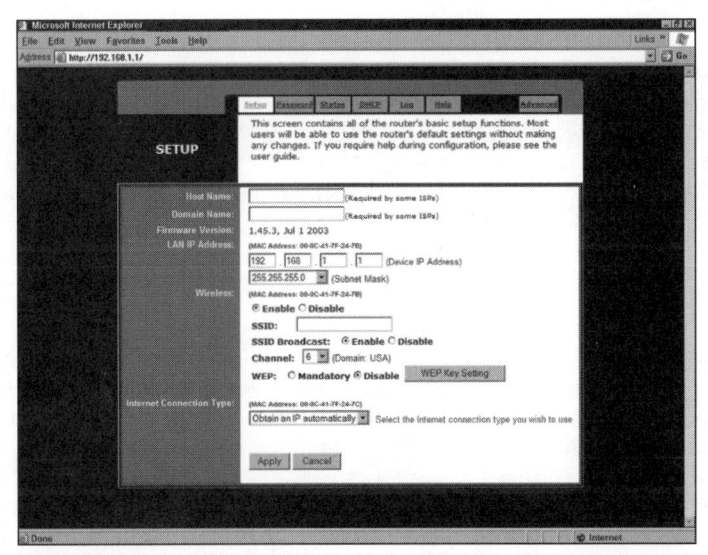

Figure 6-3: Connecting to an access point via a Web browser

Changing the default IP address settings

In Chapter 2, I explained how IP addresses work. Each device has a default IP address, and if the access point is also a router with DHCP functionality it will also have a default IP address range. The default IP address range (or subnet range) is the range of IP addresses available for *clients* using that particular access point.

It's important to change your access point's default IP address to make it more difficult for someone to connect to the device and reconfigure it. If a cracker knows which brand of access point you are using, perhaps because you failed to change the default SSID, he can try to connect to the default IP address for your device. If you are using the default username and password, the cracker can log in and take control of your access point.

Like other default settings, the default IP addresses are well known. Manufacturers, and sometimes users, usually assign a default IP address in the beginning of the network's IP address range. Many access points have a default IP address of 192.168.0.1 thru 192.168.0.5 or something similar.

The default IP address range on many devices is also well known. A cracker who knows the default IP address range can assign a valid IP address to a wireless adapter and connect to your network in the event that you have disabled DHCP. Knowing the default address range for your access point can make it easier for a cracker to discover the IP address of your access point even if you have changed it.

Cross-Reference   To read more about IP addresses see Chapter 2.

You should change the default IP address for your access point to something that isn't easy to guess. Many people assign an IP address from the first 10 or 20 addresses in the network range. If the network range is 192.168.0.1 thru 192.168.0.255, then the access point is likely to have an address between 192.168.0.1 and 192.168.0.20. Change the default IP address range, and assign an address to your access point that a cracker won't easily guess.

Avoiding DHCP

Using DHCP simplifies adding clients to your network. Unfortunately, it can also simplify a cracker connecting to your network as well. Even if you have taken the steps above, if you are using DHCP a cracker can connect to your network, and use your Internet connection.

Often, all a hacker has to do to connect to a WLAN that is using DHCP is to *associate* (connect) with the access point. Windows XP will automatically attempt to connect to a detected wireless network, and DHCP facilitates this by automatically assigning an IP address.

If you have a small network, you can assign *static* (unchanging) IP addresses to each of your network clients and then disable DHCP. This makes it more difficult for a cracker to obtain a valid IP address, though not impossible. Static IP addresses also make it easier for you to filter network traffic based on IP addresses.

Cross-Reference: Refer to Chapter 2 for more information about DHCP. See Chapter 11 for more detail about disabling DHCP and assigning static IP addresses.

Filtering network traffic

Another step you can take to make it more difficult for crackers to connect to your network is to implement *filtering* on your access point. Not all access points have this feature, but many newer access points with router functionality now include this option.

Filtering enables a router to allow or disallow connections based on the IP address or MAC address of the client attempting to connect. If a computer's MAC or IP address isn't on the allow list then the router ignores traffic originating from that computer, filtering it out and preventing that computer from connecting to the network.

Every networking device, including wireless access points and adapters, has a unique MAC address. The manufacturer encodes the MAC address at the factory, and in most cases it's permanent and users can't change it. This makes filtering based on MAC address useful and effective.

When you enable filtering on your access point, you are telling the access point which computers it can talk to based on IP address or MAC address, and that it should disregard communication from all other clients.

Cross-Reference: While filtering is effective, crackers can bypass it. Some wireless adapters allow users to change the adapter's MAC address. A cracker can listen to network traffic and, using software available on the Internet, analyze data packets to find out the MAC addresses of legitimate clients. He can then assign one of these MAC addresses to his wireless adapter and *spoof* (impersonate) a legitimate client. This and similar attacks are discussed in Chapter 4.

Using encryption

Years ago, I took a jungle survival course. When talking about the different dangers we might encounter, the instructor warned us about the numerous large jungle cats in our vicinity (he may have just been trying to scare us, but we took him at his word). When asked what we could do to protect ourselves if we encountered one of the large predators he said, "Run."

Most of us doubted our ability to outrun a large jungle cat, but when we pointed this out to our instructor he answered, "You're missing the point; you don't have to run faster than the cat, just faster than everyone else on your team." The meaning is that predators aren't going to work harder than necessary to catch a meal.

Similarly, a cracker isn't going to work harder than necessary to crack a network. The majority of home users fail to secure their wireless networks, and as a result, their WLANs are easy targets. Given the hundreds of networks that a wardriver finds, it's likely that your network will be ignored if you have taken steps to secure it. You don't have to outrun the cracker, just outrun your neighbors.

In addition to the other methods mentioned, you can enable encryption on your network to make it a less appealing target. Most WLAN equipment has Wired Equivalent Privacy (WEP) functionality built in. WEP encryption isn't perfect; in fact, most security professionals consider it extremely weak, but it's better than using no encryption at all.

Most stumbling software, such as NetStumbler, can notify a wardriver if a network has encryption or not. While there is software available that a cracker can use to defeat WEP encryption, it adds one more step and another hurdle to clear. A cracker who isn't especially proficient will probably just move along rather than bother with the obstacles.

Newer equipment has an improved encryption technology available called Wi-Fi Protected Access (WPA). WPA is a more robust encryption algorithm based upon the 802.11i security protocol released by the IEEE. Another version of WPA, known as WPA-2, is in development, and it should be available on some devices by late 2004 or early 2005.

Cross-Reference WEP and WPA are discussed in depth in Chapter 10, and their implementation is covered in Chapter 11.

Legality of wardriving

Let me start this section with a disclaimer; I'm not an attorney, nor am I an expert on criminal law regarding computer systems or communications equipment. Laws governing computer crime vary among states, and the federal government. What's legal in California may be a felony in Michigan or vice versa. It's up to you to find out what a cracker can and can't do where you live.

That said, simply driving around and passively listening for signals from access points shouldn't be against the law in most places because Wi-Fi uses public, unlicensed frequencies. Listening to Wi-Fi signals is technically legal under federal law.

Connecting to someone else's private network may be a crime in some states. Laws are evolving to address this, and attorneys are working to apply existing laws, such as those governing telecommunications, to wireless networks. The nature of Wi-Fi and the fact that operating systems like Windows XP are designed to try and automatically connect to access points complicates matters. It's not unheard of for neighbors to connect to one another's WLANs unintentionally, often without either party realizing it.

However, if someone deliberately bypasses security features in order to gain access to a private network, the chances are greater that he's broken computer trespass laws where you live. If he damages or alters those systems, steals information, or uses them to commit another illegal act, then that cracker certainly will be in hot water.

On The Web For specific legal information for individual states, visit the National Conference of State Legislatures at `www.ncsl.org/programs/lis/cip/hacklaw.htm`.

Warchalking

During the great depression, American hobos developed a language of pictographs that they used to leave messages for one another, as shown in Figure 6-4. Using these symbols, a hobo could inform other hobos passing through a rail yard or town about dangers or where they could go to find food and shelter.

Figure 6-4: Hobo symbols

Hobo pictographs inspired an activity called *warchalking*, which is related to wardriving. Warchalking is when a wardriver uses chalk to mark symbols on walls or pavement that indicate the presence of a wireless access point (see Figure 6-5). From what I can gather, few people actually engage in warchalking. At WUGs that I've visited, I have yet to find a single person that actually does this, and a recent poll on netstumbler.com had zero respondents who indicated that they have warchalked.

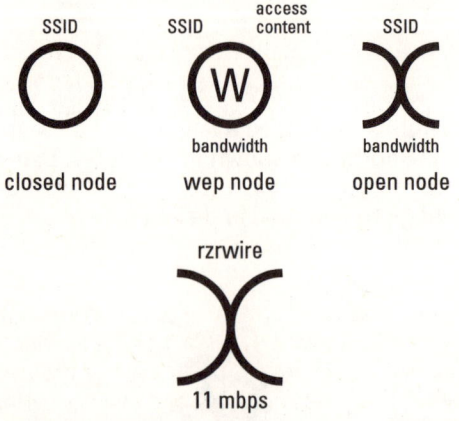

Figure 6-5: Warchalking symbols

Wardrivers can use *Global Positioning System* (GPS) devices and mapping software to create computer maps showing the locations of hundreds of discovered access points. They can upload these maps and share them on Web sites, so getting out of a car to warchalk is pointless.

For more information on warchalking, visit www.warchalking.org.

Warspying

Though not directly related to Wi-Fi networks, *warspying* targets another common wireless technology found in homes and businesses: wireless video cameras. Like wardrivers, *warspies* drive (or walk) around with wireless receivers looking for signals from wireless devices. Warspies, however, are looking for video signals, not Wi-Fi networks, and they find them. *2600 Magazine* coined the term warspying.

The most common wireless cameras are those sold by X10 Wireless Technology. You may be familiar with these cameras because of the Internet spam and pop-up advertising that the company used to promote the product. X10 cameras are completely insecure technology. If you use one of these cameras, you are broadcasting the image to the world, and anyone with an X10 receiver or similar gear can view the signal.

There is no gray area regarding the legality of intercepting the video signal from these cameras; it's legal. Wiretap laws forbid the interception of audio signals, not video. Since these cameras aren't broadcasting sound, anyone can peek. They also broadcast in the same public frequency as Wi-Fi (2.4GHz), so you can't expect privacy there.

These cameras have no security features and no built in encryption. There isn't any way you can prevent a warspy from viewing the signal. Except, maybe, if you stop using the camera. If you choose to use a wireless video camera, be careful where you point.

Software for Defense

There is plenty of software available that facilitates wardriving and wireless cracking; unfortunately, there really isn't much in the way of software that you can use to defend yourself. The best thing you can do is download a couple of the programs that wardrivers use so that you can check your own network and see how vulnerable you are.

Using a wireless laptop and a stumbling program, you can go outside and determine how far your Wi-Fi signal travels beyond the walls of your house. Using this software, you'll get an idea of what wardrivers see when they find your network. Two of these programs are NetStumbler and Kismet.

Together, these programs cover the two principal forms of scanning — active and passive. NetStumbler is an active scanner. It sends out packets and tries to get access points to answer. Kismet can scan passively by listening for Wi-Fi broadcasts and analyzing the network traffic. Kismet is more effective at discovering hidden networks (those with SSID broadcast disabled), but you'll need Linux to run it (a Windows port is in development). You can find out more about these programs on the Web sites listed at the end of this chapter.

FakeAP is a useful program for deterring wardrivers. FakeAP helps hide your access point by giving wardrivers what they are looking for. It broadcasts hundreds of thousands of fake access point signals (more than 50,000, according to the creators) hiding your access point among all of the fake ones.

FakeAP runs on Linux and takes a little skill to set up, but it's certainly an effective way to help hide your access point in plain sight. You can download FakeAP at www.blackalchemy.to/project/fakeap.

More About Wardriving

The Internet is full of information about wardriving. There are hundreds, if not thousands, of sites devoted to the hardware, software, and general how-to of locating (and compromising) access points. The following sites contain in-depth information and maintain links to other wardriving sites worth visiting. They're a good starting point for further investigation of wardriving and general 802.11x security topics.

- **World Wide WarDrive** (www.worldwidewardrive.org) — WWWD is a coordinated effort to collect statistical data about Wi-Fi use and security. The site contains useful information, and links.
- **Wardrive.net** (www.wardrive.net) — This Web site maintains links to 802.11 security-related Web sites, whitepapers, articles, how-to's, and more. It's comprehensive and well maintained.
- **Wardriving.com** (www.wardriving.com) — Lists of hardware, software, links, and other useful information about wardriving.
- **War-driver.com** (www.war-driver.com) — Wi-Fi-related security, hardware, software, links, and other useful wardriving information.
- **Warchalking.org** (www.warchalking.org) — If you're interested in learning more about warchalking, and wardriving in general, go to this site.
- **Seattle Wireless** (www.seattlewireless.net) — A good site for general Wi-Fi information and security issues.

Summary

In this chapter, I introduced wardriving, the chief threat to your home WLAN. Among the topics I presented were:

- An explanation of wardriving
- Introduction to warchalking and warspying, activities related to wardriving
- Some software that you can use to defend your WLAN
- Some useful web resources related to wardriving

Wardriving is an extremely popular activity; chances are wardrivers have been through your neighborhood at some point, and may have identified many of the wireless networks. It's important that you take the time to secure your WLAN to deter wardrivers and crackers. The material in Part 2 of this book will aid you in doing this.

✦ ✦ ✦

Viruses and Wi-Fi

CHAPTER 7

Computer viruses, worms, and other malicious programs are a major headache as well as security problem. In addition to the threats that come from the Internet, your Wi-Fi network can be exposed to viruses by other avenues. The fundamental advantage of Wi-Fi is mobility, the ability to roam within and between networks. Wi-Fi exposes your computers to viruses in a completely new way. This chapter examines viruses, worms, and other malicious code, as well as virus writers and hoaxes.

Understanding Malicious Software

Malicious software, *malicious code*, or *malware* is software created to undermine security, damage systems, steal information, make you miserable, or all of the previous.

Twenty years ago, most people working with computers were primarily worried about *computer viruses* that spread mainly via infected floppy disks and computer files. Today's threats have evolved to include *computer worms*, *Trojans*, *Bots*, and *blended threats*.

Malware can spread or attack in different ways, so it's important to familiarize yourself with the types of threats that you are likely to face in order to take proper steps to secure your systems and protect your data.

In This Chapter

Malicious software explained

Investigating computer viruses

Digging up Internet worms

Explaining blended threats

Introducing bot software

Revealing Trojan programs

Exposing spyware

Examining viruses on handheld devices

Revealing the hoax problem

Malware is a significant threat to wireless networks and devices. In addition to all of the threats that Ethernet networks face, wireless opens up new avenues of attack for malware to exploit. However, the situation isn't completely dire, and you can correct these problems.

On a wired network, the most likely vector for infection by malware is the Internet. However, on a WLAN there are additional vectors to worry about (see Figure 7-1). It's possible for malware from outside wireless sources to attack your network. Some of these sources include:

- Wardrivers and crackers
- Cross-contamination from adjacent WLANs
- Infected mobile clients

Wardrivers and crackers can purposely infect your wireless clients after accessing your WLAN. This doesn't even have to be intentional; individuals who are only looking to use your Internet connection could possibly infect your computer if they haven't taken steps to defend their own.

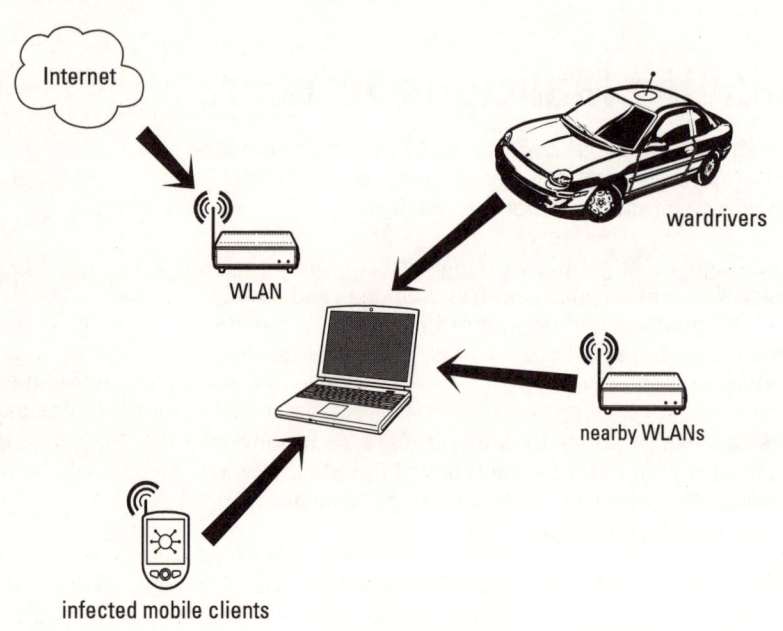

Figure 7-1: Possible vectors for infecting a WLAN

Often, neighbors may have WLANs operating on the same channel and even using the same default SSID. Because Windows XP automatically attempts to connect to detected wireless networks, a neighbor could inadvertently connect to your WLAN rather than his own, creating an avenue for infection through his computers (see Figure 7-2). You may even inadvertently connect to a neighbor's WLAN, creating the same problem in reverse.

Clients that inadvertently connect to an adjacent WLAN may be compromised by a worm or virus

An infected client from a nearby WLAN may connect to your WLAN and infect your network clients

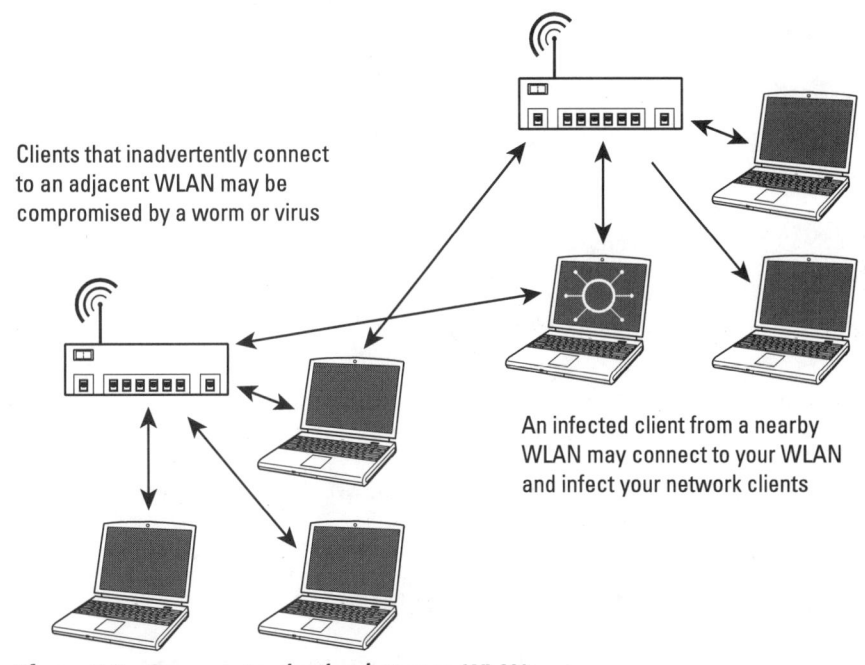

Figure 7-2: Cross-contamination between WLANs

A mobile wireless client like a laptop or PDA can also introduce an infection into your network. Although there are few viruses that affect PDAs, your PDA can carry a virus that affects your PCs. For example, if you downloaded a ZIP file or a Word document to your PDA and then transferred it to you PC, it's possible that the file could be infected with a virus that targets your PC (see Figure 7-3).

Note Even though there still aren't many viruses that target PDAs, the possibility of introducing a PC (or Mac) virus onto your network through your PDA is real. Install antivirus software on your handheld devices to detect viruses, and scan all files that you transfer to your computer from a PDA.

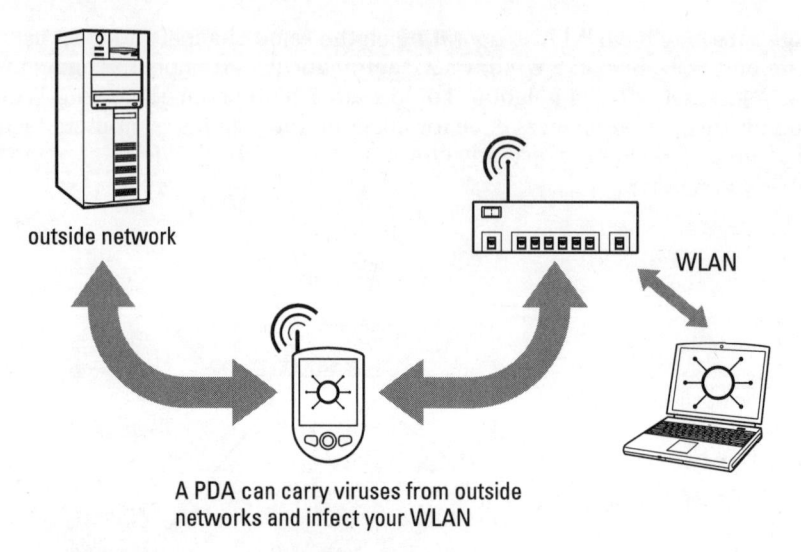

Figure 7-3: A PDA as a vector for infection

While your wireless network is vulnerable to different types of attack, you can close many of these avenues of vulnerability by taking simple steps to secure your WLAN.

> **Cross-Reference:** Read more about taking steps to secure your WLAN in Chapter 11.

Examining Computer Viruses

Computer viruses are analogous to biological viruses in many ways. A biological virus consists of nothing more than a strand of genetic code in a protein case, and it must hijack a living cell to make copies of itself. Like a biological virus, a computer virus is code that requires assistance to reproduce, but rather than hijacking a cell, a computer virus infects an executable program.

When a user launches an infected program, it activates the virus. The virus then adds itself to other programs, and in some cases may execute a secondary function or *payload* (see Figure 7-4). The payload may be a harmless joke, like a funny message on the screen, or it may destroy data by overwriting essential system files or deleting documents.

A computer virus can't spread by itself; it must infect an executable program that a user then launches. Even common e-mail *macro viruses* follow this rule, infecting executable macros in e-mail programs, which a user then activates by launching an attachment.

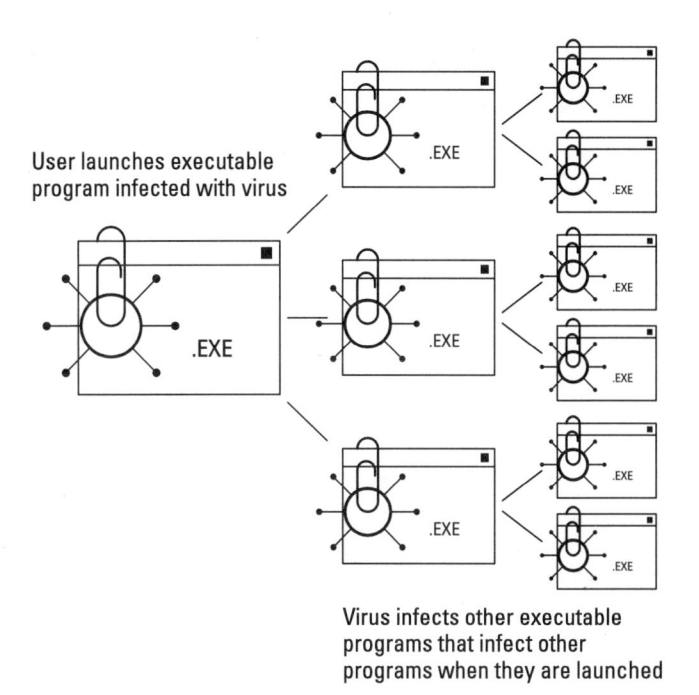

Figure 7-4: A computer virus infecting executable programs

A computer virus generally follows a simple life cycle (see Figure 7-5) that, again, is analogous to the life cycle of biological viruses. The cycle begins with its creation and ends with its identification and termination. The cycle may begin again with the modification of the virus (creation of a variant) and its subsequent release. The steps in the computer virus life cycle are:

1. Creation, or birth
2. Release, or initial distribution of the virus
3. Trigger, either a date or event (optional)
4. Activation
5. Detection
6. Elimination or removal
7. Modification

Note Other types of malware, including worms and blended threats, follow a similar cycle.

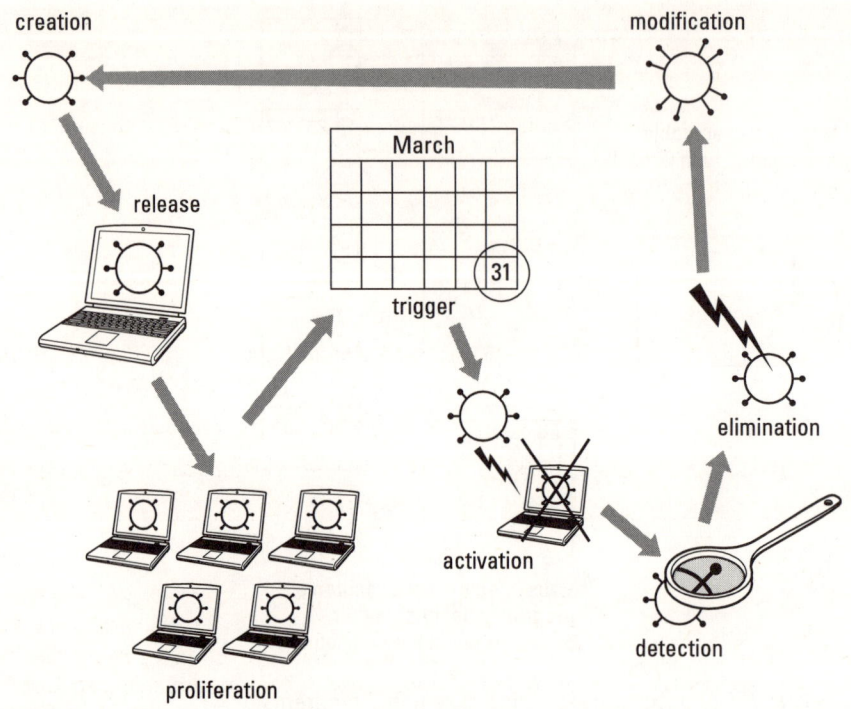

Figure 7-5: The life cycle of a computer virus

Dangerous and double extensions

Because viruses infect executable files, it's important that you don't arbitrarily execute files that you receive as e-mail attachments. Not all executable files end with the .exe file extension. Table 7-1 lists some other file extensions that indicate executable files and their associated programs.

Table 7-1
Dangerous File Extensions

File Extension	Associated Program or Function
EXE	An executable file, application, or program.
VBS	Visual Basic Script. Executable code created with Microsoft Visual Basic.
BAT	Batch file (example: autoexec.bat). Although initially created for MS DOS, batch files will still execute on newer Windows operating systems including Windows 2000 and Windows XP.

File Extension	Associated Program or Function
COM	Another executable file, program, or application.
PIF	Program Information File. A link to an executable DOS file that stores information about window settings for the DOS file.
LNK	A windows shortcut used to link to an executable file.
SCR	A Windows screen saver file.
VBE	A Visual Basic Encoded script file, similar to a VBS file. It executes in the same way.
JS	A JavaScript external file, used to contain executable JavaScript rather than embedding the script directly in a web page. Potentially dangerous.
HTA	An executable HTML application file that can be embedded on a Web page.
SHS	An executable Windows OLE (object linking and embedding) package that can act as a container for executable code.

If you encounter an attachment with an extension you don't recognize, be safe. Don't open the file. Data files don't represent a threat, because data files aren't executable. Although data files can be infected with malicious code, opening the files does not activate the code because data files are not executed like application files are. Malicious code located in a data file requires an infected executable program to execute it.

An example of this is a file-infecting macro virus. The malicious code — in this case a Microsoft Word macro — is embedded in a Word document. In order for the macro virus to execute, Microsoft Word must run the macro. This is why disabling Microsoft Word's macro feature defeats macro viruses.

Because data files are usually harmless, crackers and virus writers take advantage of this to trick people into opening executable attachments. They can use double extensions to confuse people or even hide the real extension of a file. The following file names are examples of this:

✦ Iloveyou.txt.exe

✦ account_info.doc.pif

✦ yourmessage.jpg.scr

In each of these examples, the false file extension that indicates a data file is followed by the actual file extension indicating that the file is actually an executable file. Windows allows the "." character in a file name and recognizes the last three letters following the final "." as the actual file extension. The problem is that if you have enabled the Hide extensions for known file types option, Windows will hide the actual extension and only shows the file name with the first, *false* extension. Figure 7-6 illustrates this and displays one of the preceding example files in a folder window with the known extensions hidden.

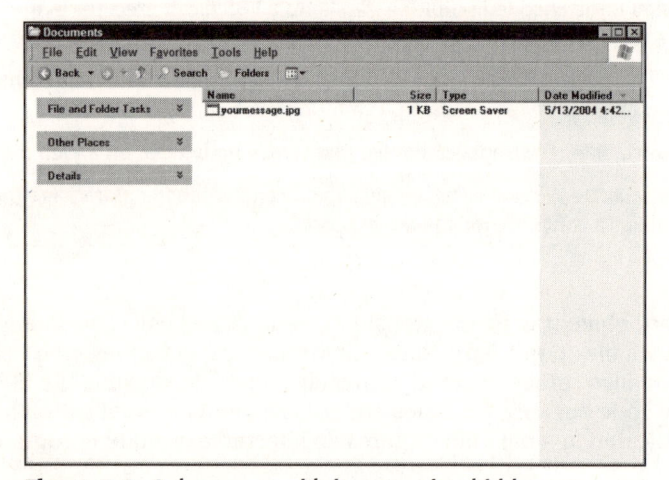

Figure 7-6: A document with its extension hidden

All of the executable files appear to be harmless data files, and an unsuspecting user is likely to launch one of them inadvertently. Another method for hiding the actual extension is to create a long file name with nonprinting characters like spaces:

yourmessage.jpg   .scr

Figure 7-7 shows this file displayed in a folder list. In the top window, the last extension is hidden because of the long file name and will remain hidden even if the system is not configured to hide known extensions. In the bottom window, you can see the actual file extension, but only by making the Name column extremely wide.

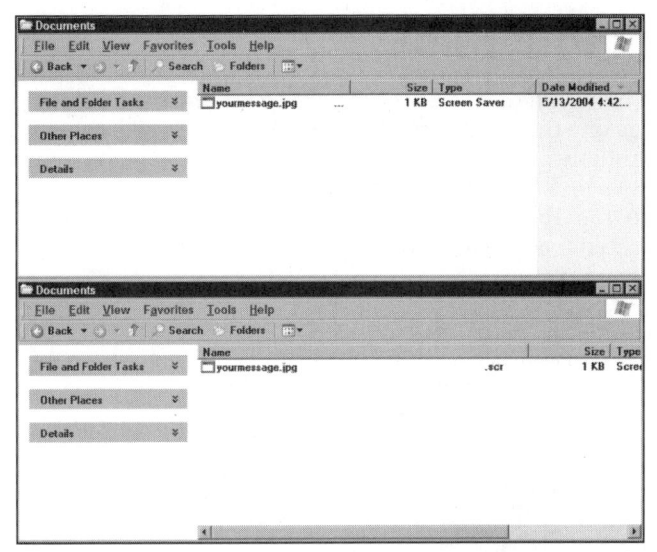

Figure 7-7: A double extension can be hidden by a long file name with blank spaces

The Microsoft fink-fund

In late 2003, Microsoft announced the creation of a $5 million antivirus reward program (now called the *fink-fund* by the security community) that provides monetary rewards to any person who provides information leading to the arrest and conviction of the creators of malicious software.

Microsoft placed the first bounties on the heads of the creators of the Sobig virus, the MSBlast worm, and the Mydoom.B worm. Each reward currently stands at $250,000. Because the creators of malware often seek bragging rights trying to prove their technical prowess, it's likely that someone knows who these people are (especially fellow crackers). Microsoft hopes that the fund will give persons with knowledge of the virus writers' identities incentive to come forward and identify them.

So far, no arrests have resulted due to the bounty placed on the heads of these people, but it's likely that this approach will aid in the investigation of future malware.

> **On The Web**
> For more information about the Microsoft antivirus reward program, visit www.microsoft.com/security/antivirus/.

Exposing Trojan Programs

Like the wooden horse of legend that the Greeks used to infiltrate the city of Troy, a *Trojan* program poses as one thing yet actually contains another. Like a virus, a Trojan requires an executable program that the user launches to initiate infection. Often, a Trojan appears to be a harmless game, utility, or program, but when a user launches the Trojan program the malicious code it contains is activated, and the computer becomes infected (see Figure 7-8).

Worm located on infected computer scans network (or Internet) to locate computers with security flaws that have not been patched

network

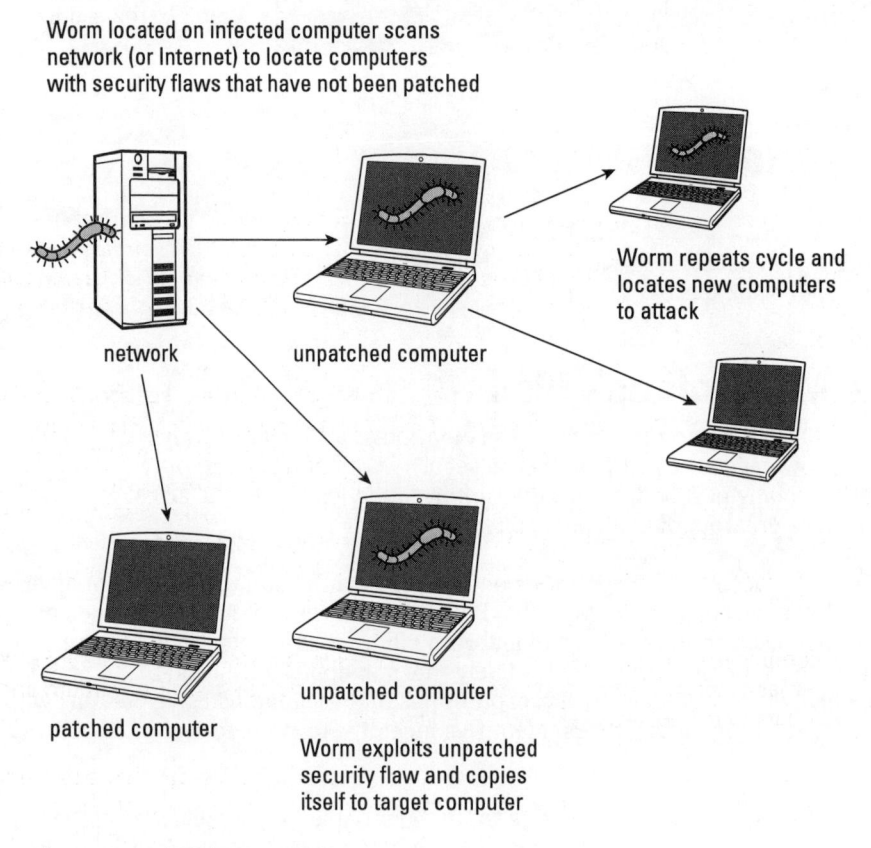

unpatched computer

Worm repeats cycle and locates new computers to attack

patched computer

unpatched computer

Worm exploits unpatched security flaw and copies itself to target computer

Figure 7-8: A Trojan program

Crackers use Trojans to spread viruses and worms, to install *backdoors* and spyware, or to steal and destroy data. The *payload* or purpose of a Trojan varies from program to program. Some computer worms create Trojan files containing copies of themselves in order to trick users into launching copies of the worm. Some worms create Trojans and copy them into shared directories of Kazaa and other peer-to-peer network services. This way, other users might download the Trojan and infect their machines.

The best way to protect yourself against Trojan programs is to avoid launching executables received in your e-mail and to install and maintain a computer antivirus program. If you download any software from the Internet, scan it to make sure it isn't infected before you install or launch it.

Discovering Internet Worms

In 1988, Robert Morris, a graduate student at Cornell University, created and released the Morris Internet Worm, a self-replicating program that spread throughout the Internet. The worm replicated out of control and crashed more than 6,000 computers (roughly 10 percent of connected computers, most of them servers), bringing the Internet to its knees.

To put the effect of the Morris Worm in perspective, the Internet was much smaller in 1988. At that time, commercialization of the Internet had just occurred, and most Internet users were at universities and government agencies. In 1988, the Internet had approximately 56,000 connected machines; by 2004, the number of machines is anyone's guess. It's simply too hard to count, but some estimates put the number of hosts at more than 231 million. Regardless of exact numbers, it is easy to see how much more of an impact worms can have on the Internet by today's standards.

> **On The Web**
> A company called the Internet Systems Consortium conducts a biannual survey of Internet hosts. For more information about the Internet Systems Consortium and the data that it collects, visit www.isc.org.

In contrast to a computer virus, a *worm* is an autonomous program that can spread across networks and replicate without user intervention. While most worm programs are malicious and designed to attack systems, researchers originally designed worms to automate useful network tasks. However, during this research it became apparent that people could also use worms to attack systems.

To access and infect computers across a network, worms exploit known vulnerabilities in applications and operating systems. First, the creator of the worm must release the program into the wild, meaning it has to be on a computer connected to a network (wireless or wired), which is in turn connected to the Internet. Once on a networked machine, the worm scans the network to locate machines with known vulnerabilities that haven't been patched to correct the flaw. This leaves those computers vulnerable to attack. When it locates a susceptible target, the worm exploits the vulnerability and copies itself into the memory and onto the hard drive of the target computer. Once it has infected a new computer it begins the cycle again and spreads to other susceptible computers (see Figure 7-9).

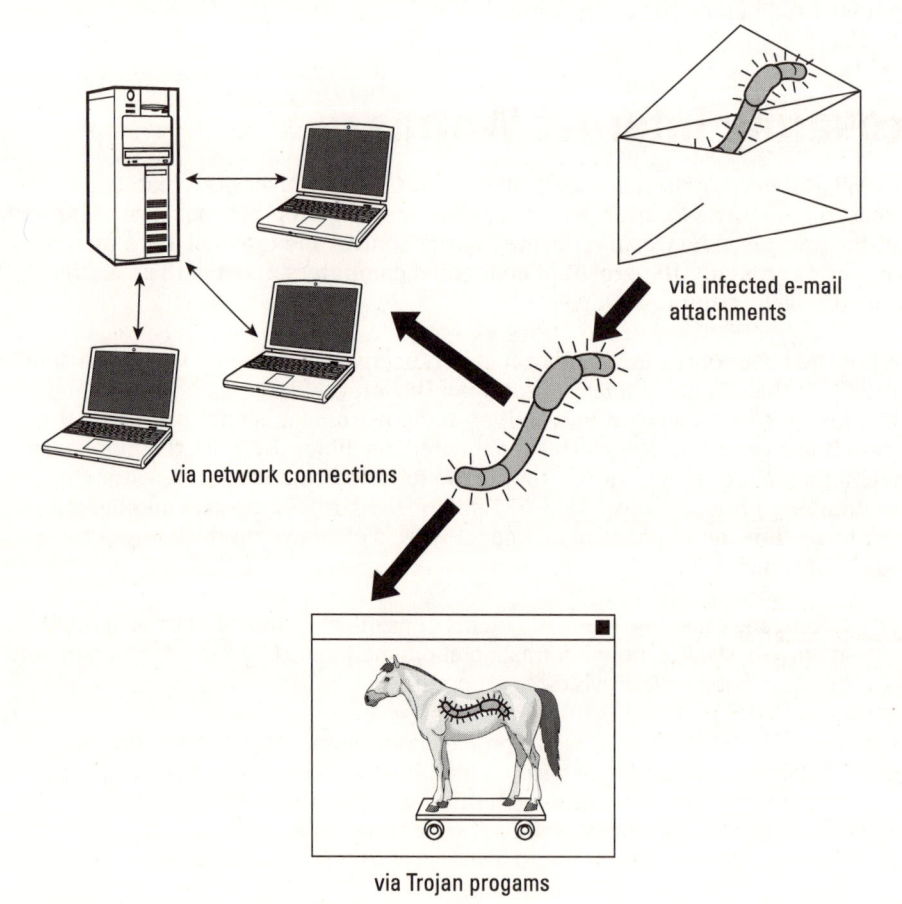

Figure 7-9: Computer worm infection cycle

Some worms spread themselves via e-mail in addition to spreading via network connections (see Figure 7-10). The worms do this by scanning the address book file of a user's e-mail program and then sending copies of themselves to the addresses they locate. Many worms can generate random subject lines and disguise the copies that they send as some other random file name the same way viruses do. When spreading via e-mail, these worms require that the recipient of an infected message launch or open the attachment containing the copy of the worm. Once the user unwittingly activates the worm, it can then spread over the network by e-mailing itself from the new computer.

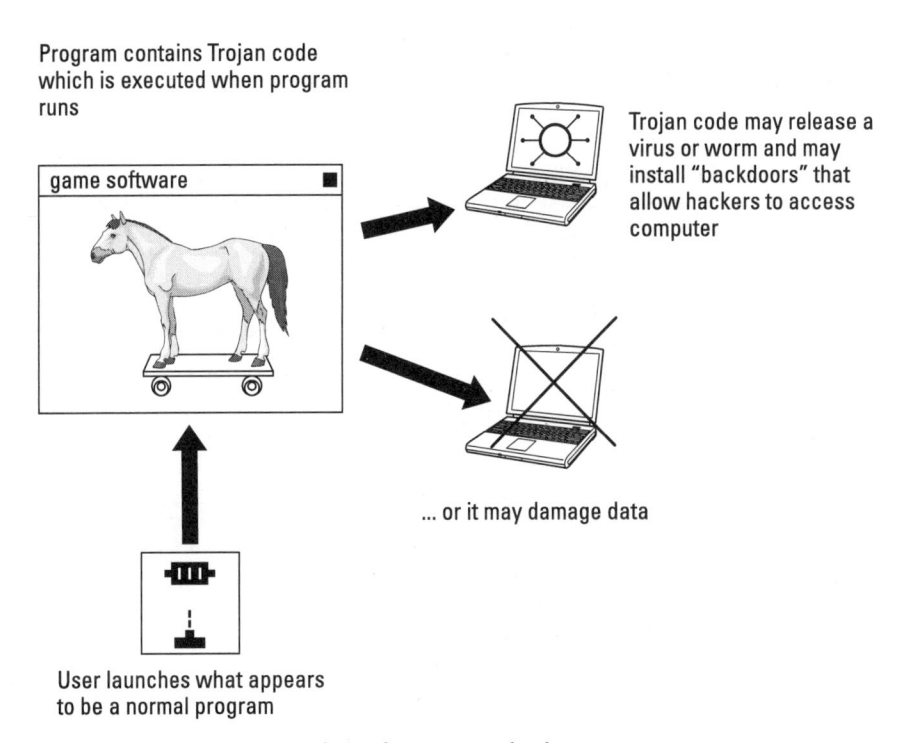

Figure 7-10: Worms spread via alternate methods

You can sufficiently protect your computer from worms by taking the following precautions:

- Install antivirus software, and update the software regularly
- Install software and operating system updates as soon as they are available
- Don't open suspicious e-mail attachments, even from people that you know

Antivirus software will detect and eliminate worms in e-mail attachments, on your hard drive, and in your computer's memory. Update your antivirus software's definitions regularly so that they remain current and can adequately defend your computer. If the software has an automatic update feature, activate it. This will make it more likely that your antivirus program will remain current as well.

Because most worms exploit vulnerabilities in applications and system software, patching these software bugs will cut off the worm's avenue of attack. Software vendors patch most vulnerabilities before crackers can figure out how to exploit them. This means that if your computer has all the available patches installed, it's less likely that a worm will infect your computer via this route.

Note As a reminder, if you reinstall your system software or applications, be sure that you also reapply all of the patches because patching will be undone when the operating system and applications return to their original prepatched state.

Just as I said for viruses, don't open suspicious e-mail attachments, even if they appear to be from someone you know. Often, an executable file can be disguised as what appears to be a harmless document; so, don't open something thinking it's safe just because it doesn't end in .exe.

Caution Crackers often attempt to trick users into launching Trojans by disguising them as system/security patches or worm/virus removal utilities. Often, they distribute them via e-mail that appears to come from Microsoft.

Microsoft does not send patches via e-mail to any of its customers—so, don't be fooled. Because there are hundreds of millions of people using Microsoft products, it is impossible to contact customers in this way. It's up to you to go to Microsoft's Web site and download updates.

Discovering Blended Threats

A *blended threat* is malicious code that uses multiple methods to propagate and attack targets. In affect, a blended threat is a worm that can use multiple automated attacks to compromise systems and propagate. Blended threats can spread rapidly and do a great deal of damage.

The *Nimda* worm, an early example of a blended threat, compromised and infected more than two million computers in less than 24 hours. Nimda spread by attacking a known vulnerability in Microsoft's Internet Information Server (IIS) software. Nimda also spread itself via infected e-mail attachments and could spread across networks by infecting vulnerable shares. Blended threats such as Nimda can cause hundreds of millions of dollars in damage.

Because blended threats can spread via multiple methods and are automated, they are more difficult to defend against than simple viruses or worms. Some of the ways that blended threats can spread are:

- Exploiting known vulnerabilities in systems and applications
- Via Web pages
- Infecting shared folders and directories
- Through e-mail attachments

Although it's an integral part of defending your computer, antivirus software alone is not sufficient to protect your systems from blended threats. For example, like many other blended threats, the CodeRed worm spread by exploiting vulnerabilities in Microsoft operating systems. Because CodeRed executed itself directly in a computer's memory rather than first copying itself to a hard disk, it was able to bypass many antivirus products.

Early blended threats infected millions of computers in a single day; newer blended threats can spread even more quickly. Before long, we will see worms that spread to millions of hosts in a matter of hours or even minutes. With the speed at which blended threats spread, it's imperative that you take steps to protect your computers. A *layered* defense, also known as *defense in depth*, is the best way to protect your WLAN. Defense in depth includes:

- Patching your system and applications regularly to eliminate vulnerabilities
- Installing antivirus software and making sure that it stays up to date
- Using a firewall to protect your network

Again, patching your system to eliminate vulnerabilities is extremely important. Many worms and blended threats have spread by exploiting *known* vulnerabilities. Vendors often release patches to correct vulnerabilities weeks prior to the appearance of malicious software designed to exploit the flaws. Because millions of users fail to download and apply these patches, however, worms and blended threats spread and do millions of dollars of damage.

Taking the time to secure your WLAN will prevent a worm from spreading to your computers from a nearby WLAN or infected computer. Recently, an associate's

WLAN was infected with the Sasser worm, even though she had taken steps to protect her network from attack via her Internet connection. She had a properly installed firewall and antivirus solution, but her WLAN was completely insecure.

A neighbor inadvertently connected to her WLAN instead of his own, and the worm spread to her WLAN from his infected computer. Had she taken steps to prevent unauthorized connection to her wireless network, the Sasser worm would have never infected her computers.

> **Cross-Reference:** Read more about how to secure your WLAN in Chapter 11.

Bot Software

An underreported threat, but one that you should take very seriously, is that which *bot software* or automated remote attack tools pose. Similar to a worm and a Trojan, bot software automatically scans for vulnerable systems and then installs itself, giving a cracker control over the system. The cracker can order the bot software to attack other computers or use it to steal information from the infected computer (see Figure 7-11).

Often, these bots run undetected, and users may be oblivious to the fact that the bot has compromised their computers. Because bots attempt to spread in a stealthier manner and don't scan huge numbers of random IP addresses while searching for targets, they don't create the amount of network noise that worms or blended threats do. This makes bots harder to detect through network traffic analysis.

Bots are becoming more sophisticated, incorporating more programmable functions so that attackers can use them for a number of different tasks. Some of these include:

- Using multiple bot-infected clients to launch denial of service (DoS) attacks on Web sites
- Using bots to spread other malicious software, including worms
- Using bots to send unsolicited commercial e-mail (spam)

> **Cross-Reference:** Antispyware software is a good idea. There are several products available that will detect a number of different bots, Trojan programs, keyloggers, and more. For more information about spyware and ways to protect yourself against it, see the section "Introducing Spyware."

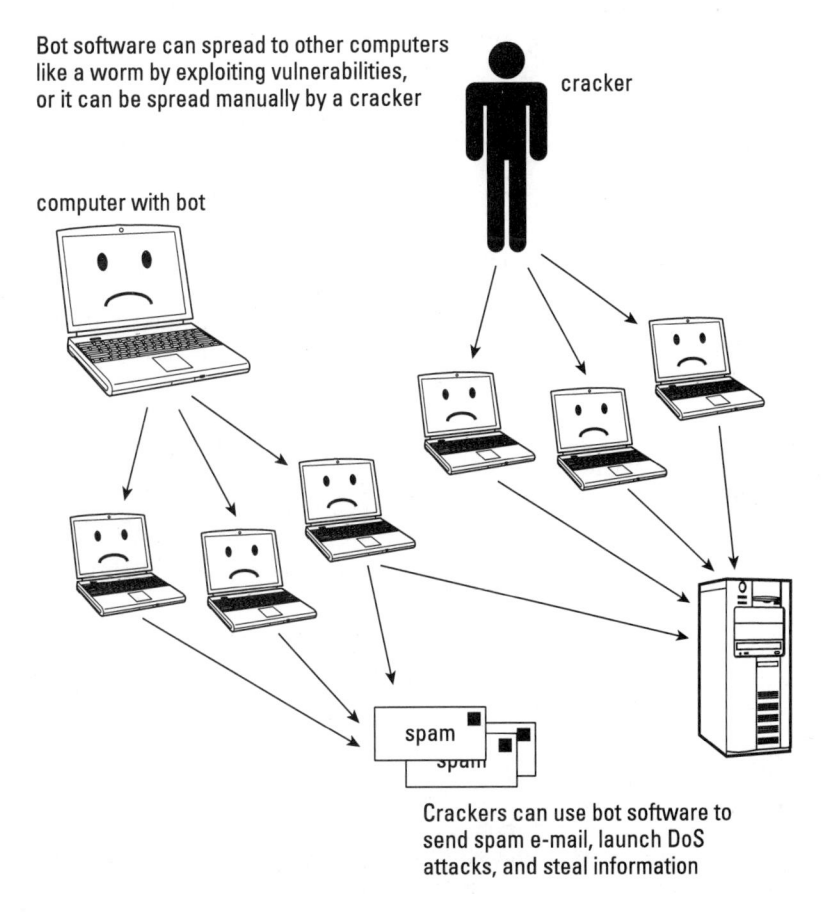

Bot software can spread to other computers like a worm by exploiting vulnerabilities, or it can be spread manually by a cracker

cracker

computer with bot

Crackers can use bot software to send spam e-mail, launch DoS attacks, and steal information

Figure 7-11: Bot software

Bot software can spread via multiple methods. Depending on the particular bot, the best strategy for protecting yourself is to do all of the following:

- ✦ Install antivirus software, and keep it updated
- ✦ Install and use a firewall
- ✦ Patch your system and applications and keep them up to date
- ✦ Install antispyware software
- ✦ Don't launch unexpected or otherwise suspect e-mail attachments

Introducing Spyware

People use the term *spyware* when referring to two different types of software. One type of spyware, also called *adware*, is used to track your Web browsing and other online habits. Advertisers and marketers sneak adware onto your computer by bundling it within other programs like games or media players.

Does this sound a lot like a Trojan to you? It should. In principal, it's the same thing, but there are two major differences. Adware doesn't destroy data or harm your system and, in most cases, advertisers get your consent before installing it. Consent? Yes, hidden within the end-user license of many programs is the disclosure that the software will collect information and report to a central tracking server.

The problem is that most people never take the time to read the end-user license of any software they install. They simply click OK and proceed with the install; creators of spyware know this. I did an informal survey of this and found one person out of 20 who claimed to have read an end-user license, and I'm sure that he thought that it was a trick question.

Protect Yourself Against Spyware and Adware

Spyware and adware are fast becoming a major problem for Internet users. Adware and spyware can hijack your Web browser, and you'll find yourself trying to Web surf against a deluge of pop-up ads, and have your browsing and searching redirected to hijack sites.

Running an anti spyware/adware application is important and will save you much grief. A good choice is Lavasoft's Ad-Aware. Ad-Aware can detect and remove most spyware and adware, as well as:

- Block installation of adware/spyware via the Web and Active X controls
- Block pop-up advertisements
- Block data-collection and mining through cookies
- Backup your registry to enable easy restoration
- Update Ad-Aware, enabling Ad-Aware to deal with new threats
- Provide third-party plug-ins that give greater ability to remove difficult adware and spyware

There are several versions of Ad-Aware available; you can download a basic free version that will clean your system, removing adware and spyware. To block the installation of new spyware and adware, you must upgrade to Ad-Aware Plus. All versions of Ad-Aware are available for download at www.lavasoft.com.

Almost any program can be adware: product demos, shareware, free software, and even shrink-wrapped software that you buy in a store. Many software developers are selling advertising within their programs and bundle the adware to earn additional income from their product. The problem with adware is that it can use up your Internet bandwidth, and you have no way of knowing exactly what information the software is gathering.

The second and most dangerous type of spyware includes applications called system (or Internet) monitors and keyloggers. These programs monitor all of a user's activity on a computer and report it back to a cracker or some other unscrupulous individual. A keylogger monitors every key typed at the keyboard and logs it into a file for later viewing. The software may also e-mail the log to a third party.

Companies market keyloggers and system monitors to jealous spouses and to employers who want to know what their employees are doing on the Internet. There are many of these programs for sale, so it doesn't take a cracker to find and install one. These programs are simple to set up and use, and some of them can even install remotely, without the victim even knowing (again, like a Trojan).

Spyware is usually legal as long as you are installing it on your own computer and not targeting someone else. Installing a monitor or keylogger on someone else's computer without that person's knowledge could amount to information theft.

To protect yourself against spyware, you should take the following precautions:

- Keep your antivirus software up to date.

- Download and use a *spycleaner*, which is a software that identifies and removes spyware.

- Don't use software that is supported by advertising.

- Install a personal firewall that monitors outgoing connections Antivirus software will identify some keyloggers and system monitors, especially those programs that crackers frequently use. Antispyware software can identify and delete spyware (and adware). Finally, a personal firewall will alert you if an application on your computer is trying to access your Internet connection and phone home.

Columnist and programmer Steve Gibson has created a free online spyware detector and remover called OptOut. To try OptOut and check your system, visit http://grc.com/optout.htm.

Investigating Viruses on Handhelds

Many people mistakenly believe that PDAs and other handhelds are safe from computer viruses and malware; this isn't true. Granted, there has been very little malware created to target handheld computers, but this is changing, as the installed base of these devices gets larger.

Virus creators want more bang for their buck, which is why they spend more energy targeting systems with a large market share or high profile. This is why there are far fewer viruses that attack Macs than there are that attack Windows machines. Because there are so few Macs as compared to Windows machines, few virus writers want to waste their time going after them. There may not be too many viruses affecting them now, but as the number of PDAs increases, they will begin to get more attention from virus creators.

There are antivirus programs available for Palm and PocketPC machines. The best of these not only target known handheld viruses, but also will detect and remove viruses that affect PCs as well. This is important because your handheld can easily transport a PC virus and infect your WLAN. If you dock your handheld to your PC or connect to a network, either Ethernet or Wi-Fi, you're vulnerable to viruses transported by a handheld. Table 7-2 lists some of the antivirus software available for handhelds.

Table 7-2
Antivirus for Handhelds

Product Name	Handheld Platform	Developer and Web Site
Airscanner Mobile AntiVirus Pro	Pocket PC	Airscanner Corp. www.airscanner.com
VirusGuard	Palm OS	Mevix Software, can be downloaded at www.handango.com
BitDefender	Palm OS	BitDefender, www.bitdefender.com
MindSoft Pocket VirusShield	Pocket PC	MindSoft, www.mindsoft.com
F-Secure Anti-Virus for Pocket PC	Pocket PC	F-secure, www.f-secure.com

Understanding the Hoax Problem

In addition to actual viruses, you have to contend with virus hoaxes, those alarmist e-mail warnings about the newly discovered or most devastating virus ever found. There are hundreds, if not thousands, of these hoaxes circulating around the Internet at any given time.

Hoaxes are a serious problem. Users waste time and bandwidth when they forward virus hoaxes to friends and family. Hoaxes also spread misinformation, and a cracker can use a hoax to goad a user into making a change to his system or even running a Trojan program posing as a worm remover or system patch.

A Trojan program may also arrive in your e-mail inbox attached to a virus hoax. To protect yourself, the best thing you can do is to take the time to educate yourself and learn how to identify a hoax and where you can go to look up the facts.

Identifying a hoax

Most hoaxes have a few traits in common that should tip you off that something isn't kosher. Some traits to watch for include:

- ✦ Claims to be from an authority or credible expert
- ✦ Urges you to forward the e-mail to everyone you know
- ✦ Uses technical sounding language to give it credibility

The following is an example of a virus hoax called the good times virus, and it demonstrates some of these attributes. When it first hit the Internet in 1994, panicked users forwarded so many copies of this hoax that mail servers crashed at many companies and ISPs. Even network administrators fell for this and forwarded it to all of their own personnel.

```
The FCC released a warning last Wednesday concerning a matter
of major importance to any regular user of the InterNet.
```

First tip-off, the hoax uses the FCC for credibility. The FCC has nothing to do with computer viruses or security. If an agency actually did release a warning it should have a link to the organization's Web page containing the actual info.

```
Apparently, a new computer virus has been engineered by a user
of America Online that is unparalleled in its destructive
capability. Other, more well-known viruses such as Stoned,
Airwolf and Michaelangelo, pale in comparison to the prospects
of this newest creation by a  warped mentality.
What makes this virus so terrifying, said the FCC, is the fact
that no program needs to be exchanged for a new computer to be
infected. It can be spread through the existing e-mail systems
of the InterNet. Once a computer is infected, one of several
things can happen. If the computer contains a hard drive, that
will most likely be destroyed. If the program is not stopped,
the computer's processor will be placed in an nth-complexity
infinite binary loop - which can severely damage the processor
if left running that way too long. Unfortunately, most novice
computer users will not realize what is happening until it is
far too late.
```

Plenty of nightmarish, technical-sounding horrors to go around there, never mind the fact that it's all gibberish. Other versions of this hoax also urged people to send the e-mail to everyone they knew.

On The Web: If you receive a virus warning, the best thing you can do is research it before you send it to anyone. There are two great sites that I recommend for this purpose.

The first is `www.vmyths.com`. This is a great site that debunks all of the hysteria surrounding computer viruses. Best of all, this site actually is maintained by an expert.

The second site is `http://hoaxbusters.ciac.org` maintained by the Computer Incident Advisory Capability (CIAC) of the Department of Energy. Hoaxbusters is a clearinghouse of information about hoaxes and isn't limited to virus hoaxes.

Summary

Armed with this information, you're now well on your way to protecting your WLAN from malicious code. The topics I presented in this chapter include:

- Introduction to malicious software
- Explanation of computer viruses, Internet worms, blended threats, and Trojan programs
- The threat from bot software, and spyware
- The emerging problem of viruses on handheld devices
- The problems caused by virus hoaxes

The days when you could leave your antivirus software turned off, or worse, not install any at all, are long gone. The pace of malware development is quickening, not slowing down, and the sophistication of the attacks is increasing. These two trends emphasize the need to be vigilant and to secure your computers.

✦ ✦ ✦

Protecting Yourself

PART

II

In This Part

Chapter 8
Technical Pitfalls and Solutions

Chapter 9
Wireless Privacy Concerns

Chapter 10
Encryption and Wi-Fi

Chapter 11
Securing Your WLAN

Chapter 12
Protecting Your Wi-Fi Data

Technical Pitfalls and Solutions

CHAPTER 8

In This Chapter

Common Wi-Fi problems

Detecting and addressing interference

Explaining interoperability problems

Facts about antennas

Optimizing WLAN and computer equipment

When it comes to networking, eliminating wires doesn't necessarily eliminate problems. A number of things can affect the performance of your Wi-Fi network. In this chapter, I address some of the more common ones and the methods for dealing with them.

Discovering Common Wi-Fi Problems

Many of the more common problems in WLANs are the result of interference or configuration errors. Included in this group are:

- Multipath signal propagation
- Channel overlapping
- RF interference

Explaining multipath interference problems

Multipath interference or multipath signal propagation occurs when a signal reflects off objects in its environment, causing it to arrive at a receiving antenna from more than one direction at varying times and strengths (see Figure 8-1). The multiple signals can confuse the receiving device and cause interference.

Signals reflecting off objects in the environment take multiple paths to receiving antenna

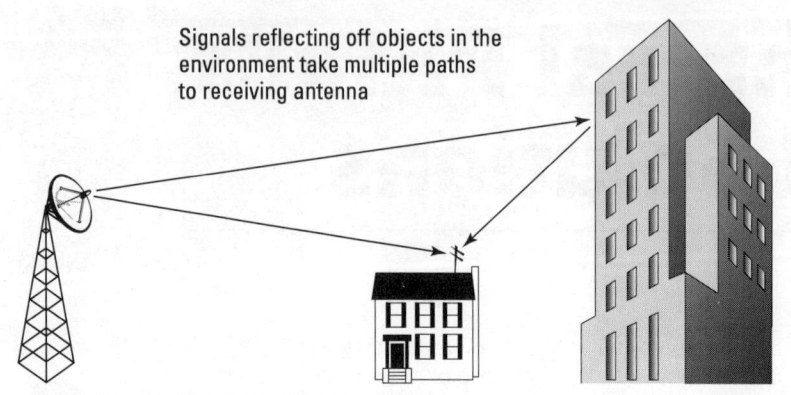

Figure 8-1: Multipath signal propagation

The interference is due, in part, to interaction of the arriving radio waves. When identical waveforms arrive at the receiver at slightly different times they can cancel each other out, which creates interference and results in lost signals (see Figure 8-2).

Identical signals arriving at slightly different times can cause signal distortion when the waveforms combine

Figure 8-2: Multipath waveforms combining

If your TV receives its signal through anan antenna, rather than cable or a digital satellite service, you may occasionally see *ghost* or doubled images. This is a visual example of multipath interference (see Figure 8-3). The television signal reflects off objects in the environment, and as a result, the signal takes multiple routes to the TV receiver. Signals arriving at slightly different times result in the ghost images.

Engineers can count the *scan lines* on a television image and based on the offset of the ghost image, can determine the distance to the objects creating the multipath interference.

Figure 8-3: Example of a ghost image

In a WLAN environment, many objects may contribute to multipath propagation. These objects include (see Figure 8-4):

- Solid, load-bearing walls, ceilings, and floors
- Metal furniture and cabinets
- Construction materials

If you suspect you are having problems with multipath interference, one way to deal with it is to use a diversity antenna system. A diversity antenna system uses dual antennas that enable the access point or adapter to get cleaner reception and avoid multipath problems.

When using a diversity antenna system, an access point doesn't simply select the antenna with the strongest signal (an often-repeated misconception). In most Wi-Fi diversity antenna systems the antennas are mounted a few inches apart, often on either side of the access point (see Figure 8-5). The signal strength difference between the individual antennas is likely to be negligible.

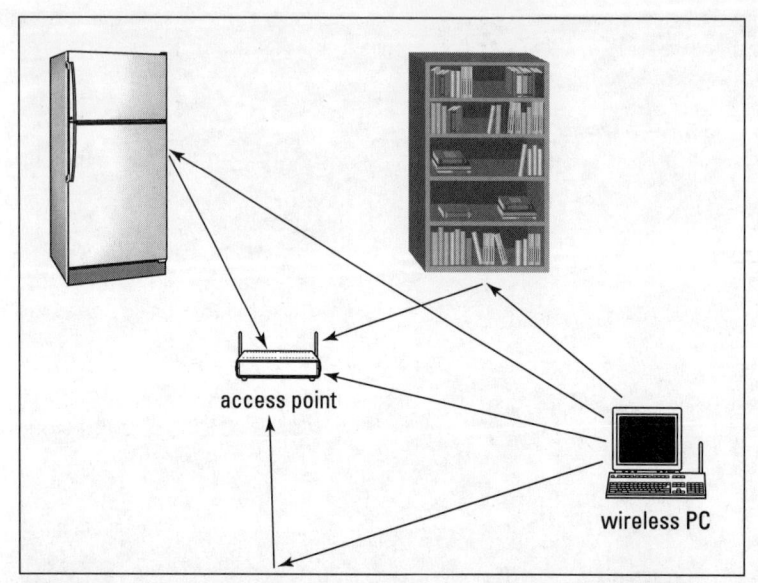

Figure 8-4: Objects causing multipath propagation in the WLAN environment

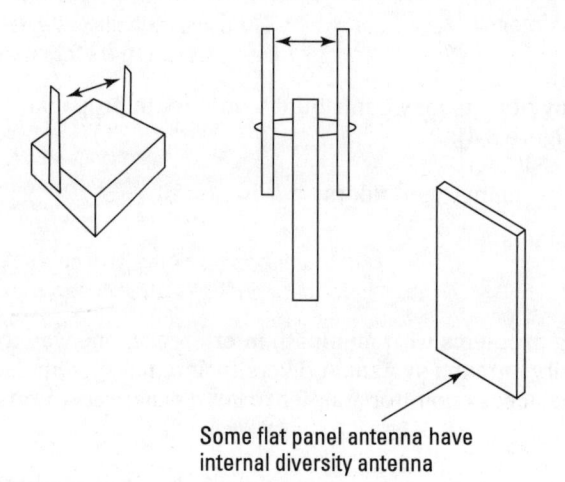

Some flat panel antenna have internal diversity antenna

Figure 8-5: A diversity antenna system

The transceiver inside newer access points or adapters can compare and discern the difference between the signals received from either antenna. By using sophisticated technological techniques, it can filter the multipath interference and eliminate the problem.

Newer Wi-Fi standards are less susceptible to multipath problems than 802.11b systems are. The reason is that 802.11a and 802.11g use a broadcast technology called *Orthogonal Frequency Division Multiplexing* (OFDM), which is less susceptible to multipath problems. 802.11b uses *Direct Sequence Spread Spectrum* (DSSS) technology, which transmits data in a wider spectrum than OFDM. Portions of a DSSS signal reflect off objects differently, allowing for more chances for multipath propagation.

Another way to identify and prevent multipath and other WLAN issues is to conduct a site survey. A site survey is an in-depth study of the environment in which your WLAN will be operating. Its purpose is to identify potential problems and plan around them.

Intentional versus Inadvertent

Theft of service may not always be intentional. Other wireless users may inadvertently connect to your network without even realizing they have done so. In areas where neighbors are physically close, such as apartment buildings, it's becoming common for people to accidentally connect to another's network, especially if they are using default settings on their equipment. In addition, Windows XP automatically identifies and attempts to connect to access points, which makes inadvertent connections more likely.

As an example, a friend of mine was having trouble setting up his WLAN. He didn't realize that his neighbor also had a WLAN and was using an access point from the same manufacturer. In addition to this, they were both using the default SSID (Linksys) and the default network settings.

The signal from his neighbor's WLAN was stronger, and his client adapter automatically connected to it. For months, he was using his neighbor's broadband connection without even realizing it. This went on undiscovered until the neighbor went on vacation and turned off his gear. Suddenly, my friend lost his Internet connection.

Until that point, he had no idea that he was using someone else's network and Internet connection. Technically this was theft of service, but it certainly wasn't intentional.

If you've taken the steps to make your network reasonably secure, you probably won't have to worry about theft of service.

A full-blown site survey probably isn't necessary for a WLAN that's going to be located in a home or small office, but you should still take the time to plan the layout of your network in advance so that you can minimize problems that you may encounter when you install your network.

Adjacent channels overlapping

802.11b and 802.11g equipment broadcast data using one of 11 different channels within the 2.4 GHz frequency band. 802.11a uses eight channels (see Figure 8-6). As the figure illustrates, the frequency band of each channel overlaps. In other words, each channel shares some of its frequency band with adjacent channels.

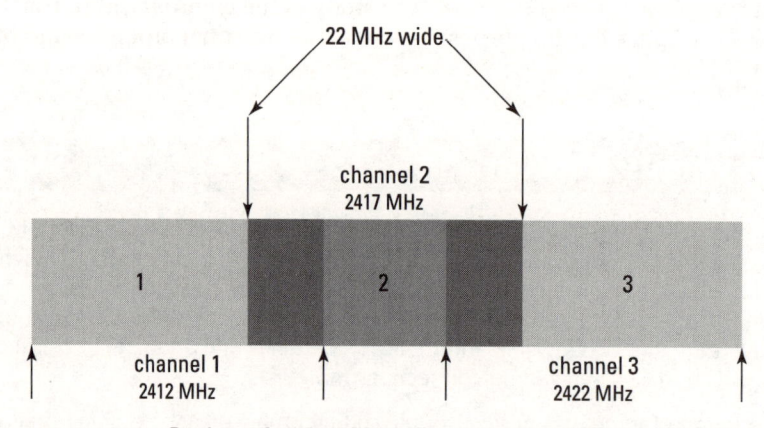

Figure 8-6: Wi-Fi channels

Because of this overlap between adjacent channels, you should only use channels 1, 6, or 11 for your equipment in order to avoid problems with nearby WLANs. Most manufacturers set their access points to one of these channels by default. Because of this, these are also the channels that your neighbors are most likely to use. If you run into interference from another network try choosing a different channel to minimize interference problems.

Identifying theft of service

One problem with wireless networking is that anyone can receive your signal, not just the people you authorize. It's possible that neighbors or passersby could connect to your WLAN and use your Internet connection without your consent. The result could be slower Internet access for you as other people use more of your bandwidth. In some cases, you may find yourself in hot water with your ISP.

Because wireless networking has made sharing a single Internet connection — and bill — even easier, some ISPs are lobbying congress and the FCC to prevent their customers from sharing bandwidth with other users. This is partly in reaction to customers setting up hotspots and sharing service with their neighbors.

In many cases, your agreement with your ISP will specifically prohibit you from sharing service with others. Violation of the agreement constitutes theft of service, and you could lose your service or have your bandwidth reduced. It may not matter that you didn't intentionally share the connection; your ISP may decide to take action anyway.

> **Note** If your Internet connection is co-opted by the neighbors, your neighbors may be breaking the law in many states by doing so. Most states have computer crime statutes that prohibit theft of service (in many cases a misdemeanor), and existing precedents support the inclusion of theft of Wi-Fi service under these laws.

Theft of service is hard to identify because an unauthorized user can connect without you ever realizing it. Unless your network traffic shows a dramatic increase, you may never notice that it was occurring.

One way that you can identify unauthorized connections, both inadvertent and intentional, is by using a network port scanner (see Figure 8-7). A port scanner is a software tool that you can use to scan your network and identify clients by IP address. For example, if you have three computers on your network and the port scanner identifies a fourth IP address with hosts connected, then you may have someone else connected to your WLAN.

There are many port scanners available for download on the Internet. Using a port scanner takes a certain amount of network knowledge and skill, so if you decide to experiment, take the time to read all documentation and familiarize yourself with all of the available options. Don't worry about damaging your network. A port scanner generates network traffic, essentially knocking on doors and waiting for an answer. You won't harm anything, but your firewall may generate alerts due to the activity.

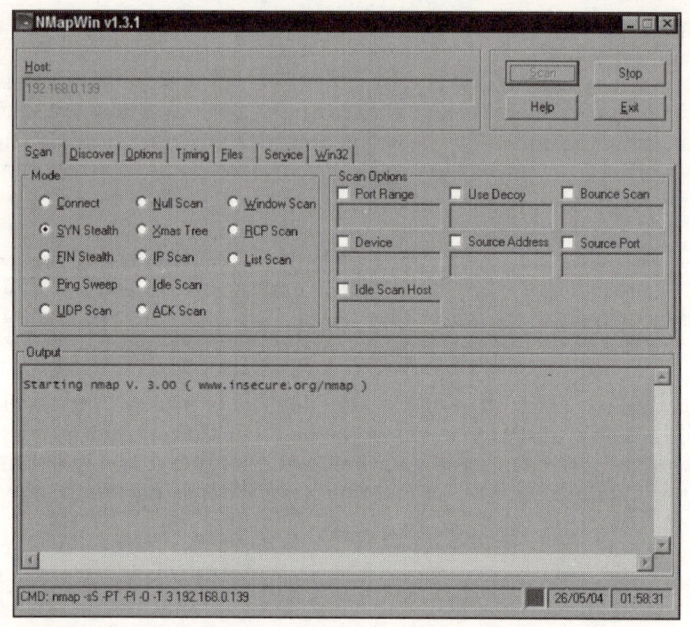

Figure 8-7: The Windows version of the NMAP port scanner

NMAP, a good port scanner, is available for download at www.insecure.org.

Read more about securing you network in Chapter 11.

Uncovering configuration errors

Occasionally, you may inadvertently make a mistake when setting up your WLAN hardware. Most often, configuration errors will keep your devices from operating at all, but it is possible that the source of your network problems could be something as simple as selecting the wrong channel on your access point.

Recently, a friend was having trouble with her home WLAN. Her access point would intermittently become unreachable through its Web interface. Everything appeared to be in order; all the clients were communicating with no problems. Eventually, the problem was traced to the way she was sharing her Internet connection.

She was using Windows XP *Internet Connection Sharing* (ICS), which has a built-in DHCP server. The XP version of ICS doesn't allow you to configure or disable the DHCP component. She had assigned a fixed address to her access point, and although it could communicate with the clients, every time the ICS computer rebooted, it became unreachable through the Web interface. In this case, the solution was to disable ICS and connect her access point directly to her broadband modem.

If you are experiencing problems, check all of your settings before you jump to conclusions and assume that you're having interference problems. Some places you may find configuration errors are:

- Incorrect antenna installation
- Incorrect SSID
- Using the wrong channel
- Adapters in ad hoc rather than infrastructure mode
- Adapter set for the wrong network
- IP address conflicts
- Connection speed set too high

Make sure that directional antennas are facing in the correct direction and that the antenna cables you are using aren't excessively long. Long cables can weaken the signal before it even reaches the antenna.

If you choose to assign manual IP addresses instead of using DHCP, make sure that you don't accidentally assign the same IP address to two different devices. Also, check the SSID for typos and to ensure that your adapters are set to connect to your network.

Excess clients may also cause performance problems on your WLAN. While some equipment advertised for home and small-office use may claim to handle up to 100 clients, in practice this claim is ridiculous. I have seen consumer access points slow to a crawl with less than 10 clients connected, especially if there is heavy network traffic. On my own network, throughput drops significantly when four or more computers are in use or when my kids are gaming over the network.

If you are going to have several clients that will be using the network heavily, I suggest investing in more than one access point. In any case, it will likely improve coverage, as well as performance.

Detecting and Dealing with Interference

As already mentioned, many devices and objects in your home can interfere with the normal operation of a WLAN. Some consumer electronics may create RF interference in the same frequency as your Wi-Fi equipment, and objects may block or reflect signals, limiting the strength and range of your signal.

Because the 2.4 and 5 GHz frequency bands are unregulated, manufacturers are free to develop devices that operate in these bands and, at the same time, are not obligated to make sure that they don't interfere with Wi-Fi signals. As a result, all of these devices are competing for the same RF real estate as your WLAN is.

Some common devices that can interfere with your WLAN are (see Figure 8-8):

- 2.4 and 5 GHz cordless phones
- Wireless speakers
- Wireless cameras
- Walkie-talkies
- Microwave ovens

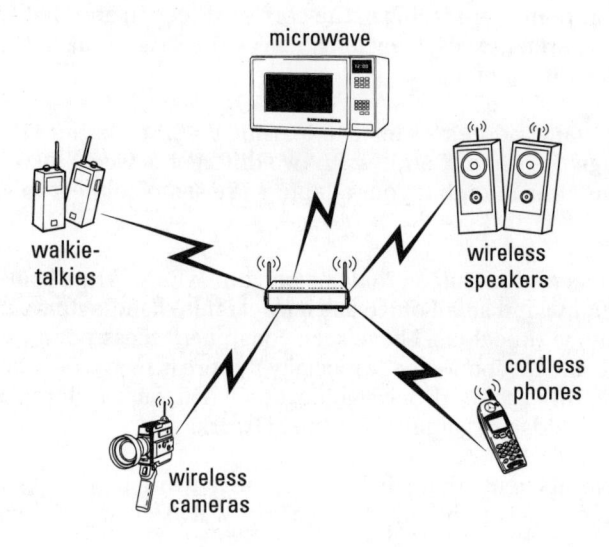

Many wireless devices can interfere with Wi-Fi signals

Figure 8-8: Sources of Wi-Fi interference

Identifying interference in your home or office

The easiest way to determine if one of these devices is the culprit of interference is to determine if your problems are intermittent and are occurring only when the device is operating or is in close proximity to your Wi-Fi devices. If this is the case, the device in question is probably causing the problem.

For example, I have a 2.4 GHz cordless phone. While the phone is charging on the base station, I don't have any problems. When the phone is off the base station and in use it interferes with the operation of my WLAN. The phone and the network are both communicating on the same frequency, and the competing signals cause interference. On the network, this results in a slowdown; on my phone, I can hear a loud clicking from the WLAN signal.

To resolve this, I changed the channel on the phone, which minimized the problem. However, not every phone or device has multiple channels to choose from, and even when they do, the channels available may still overlap with your Wi-Fi channels. Another problem is that these devices (phones, speakers, walkie-talkies, etc.) usually have considerably stronger signals than Wi-Fi devices.

The FCC sets regulations that limit the signal strength of devices on your WLAN. Wi-Fi devices usually have power output measured in hundreds of milliwatts (thousandths of a watt). The base stations of most cordless phones have a signal strength of more than 4 watts. This is far stronger than your access point or WLAN client adapters.

You may be able to position an interfering device far enough away from your access point and clients to limit the impact it has on network performance. If that doesn't help or if changing channels isn't an option (or isn't effective) there is little else you can do other than limit the use of the offending device, or replace it with one that operates at a different frequency than your WLAN.

Identifying interference outside your home or office

Identifying sources of Wi-Fi interference isn't an easy task, especially if they are originating outside your home. If you live in an apartment or dorm, where you are in close proximity to your neighbors, devices belonging to your neighbors could possibly create problems for you and vice versa.

If your neighbor has a WLAN, it may cause connection problems for your clients if your access point channel overlaps your neighbor's channel. In any case, diplomacy is probably your best bet. Try explaining the situation and perhaps offer to try to solve both of your problems. Chances are your neighbor has also noticed the interference.

Other than trial and error, there are a couple of methods that you can use to attempt to identify sources of RF interference. If you suspect that the problem is due to an adjacent WLAN, you can use the same stumbling software that wardrivers use to detect any WLANs operating in close proximity to your access point.

There are also handheld RF detectors that professionals use to detect 2.4 GHz signals when they conduct site surveys. Unfortunately, these are very expensive, often costing thousands of dollars; so unless you have access to one at work or through a friend, it probably isn't an option.

Some people recommend using a handheld *electromagnetic field* (EMF) detector to locate sources of Wi-Fi interference. Many electronics retailers sell these, and they usually cost less than $100. These detectors are standard issue among *ghost hunters*; yes, you read that correctly, ghost hunters, as in floating apparitions that say "Boo!" You can also find these devices on many ghost hunting Web sites.

The problem with using these detectors for locating Wi-Fi interference is that they detect EMF signals at very low frequencies compared to Wi-Fi. Often, EMF detectors won't detect a signal higher than 100 Hz, which is far below the 2.4 GHz frequency used in WLANS. It's possible to modify these devices so that they detect higher signals, but no modifications exist (that I could find) that will enable them to detect possible RF sources that would be of concern in a WLAN environment.

Low-frequency signals in the Hertz range do not interfere with your WLAN. If they did, Wi-Fi would never work. The electrical wiring in your home produces electromagnetic fields that are normally in the 50 to 70 Hz range, yet these don't present a problem for wireless networks either.

Note Unshielded electrical wiring and outlets may not be a problem for Wi-Fi devices, but they can interfere with PC monitors. If the EMF from an outlet is especially strong it can damage or cause distortion in an adjacent monitor.

Avoiding physical barriers

Many objects in your home, as well as the materials used in its construction, can reflect or impede Wi-Fi signals. Solid concrete load-bearing walls can block signals, as can metal furniture and appliances such as refrigerators. The mesh used in exterior stucco walls can also interfere with the WLAN signal and may impede extending your WLAN to your porch or deck. Reflection of signals off these objects can also contribute to multipath problems and limit the performance of your network.

Perhaps the easiest way to deal with these issues is through careful planning. While a full site survey of your home may be overkill, you can create a floor plan of your home or office and identify barriers to your Wi-Fi signal (see Figure 8-9). You can then position access points and clients so that you can avoid these obstacles.

Locate your access points near openings between rooms or next to windows if you're extending service to your porch or deck. Mount your access point close to the ceiling when possible to improve signal range, and avoid low obstacles such as desks and even people.

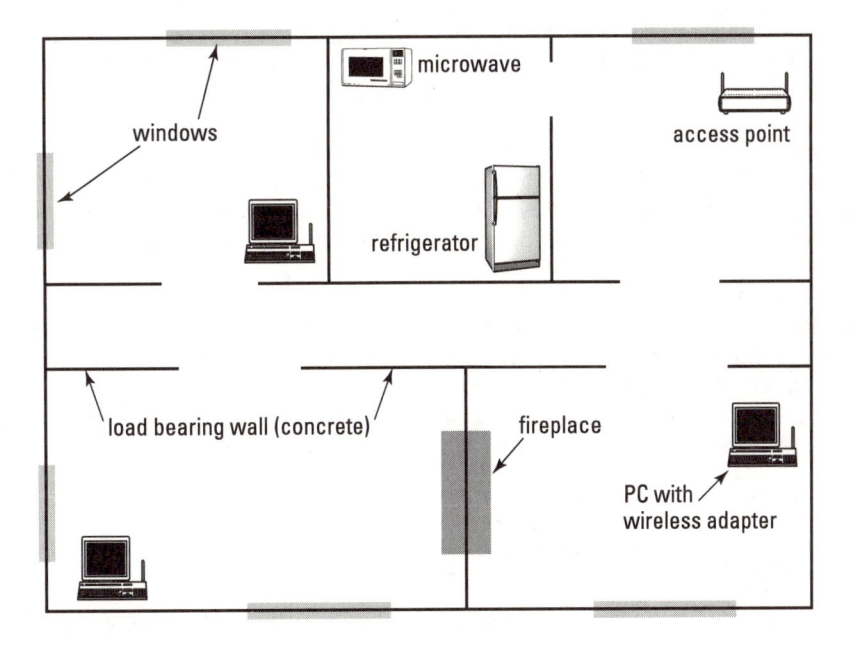

Figure 8-9: Consider making a floor plan for network planning

Defining Interoperability Issues

The newer 802.11g standard is backward compatible with 802.11b. This allows you to upgrade an existing 802.11b network incrementally, buying faster 802.11g equipment as you can afford it. However, to get the best performance out of 802.11g clients, you must have an 802.11g access point.

What manufacturers don't advertise is that while 802.11g products will interoperate with 802.11b devices, 802.11g devices will not perform as well in a mixed standard WLAN as they will in a WLAN that only includes 802.11g devices. In order for your 802.11g devices to operate at their best speed, all devices on your network should support that standard.

Note Manufacturers are beginning to address this problem, and newer access points will have fewer problems in a mixed environment; but optimal performance will always require an all-802.11g WLAN.

Unlike 802.11b and 802.11g, 802.11a devices don't operate in the 2.4 GHz frequency band. They use the 5 GHz band instead. Because of this, 802.11a devices won't interoperate with 802.11b and 802.11g devices.

There are some new access points that will interoperate with Wi-Fi clients using any of the three standards. To do this, these devices incorporate multiple radios into an access point, usually an 802.11g and an 802.11a transceiver. In effect, each of these devices is actually two access points incorporated into one box and sharing the same SSID, IP address, and so on.

Because of the extra hardware and processing power, multifunction or multiband access points are usually much more expensive than either 802.11g or 802.11a devices. Costs sometimes run almost double what you would pay for a single-band device. However, if you need to mix 802.11a, 802.11g, and 802.11b devices on the same network, a multi-band access point is your best solution.

Learning About Antennas

One of the most important components in any Wi-Fi setup is the antenna. The antenna can also be the source for many potential problems. In order to identify and correct some of these problems, it helps to have a basic understanding of what an antenna is and how it works. Antenna theory can be confusing, and frankly, very boring, so I'll keep this explanation short and as simple as I can.

Devices in your WLAN use electromagnetic waves in the radio frequency (RF) band to transmit data. Access points and adapters are actually two-way radios, or *transceivers*, that can both transmit and receive RF signals. Transceivers use antennas to broadcast the RF signal, increase signal range, and improve reception.

The chips in a transceiver create an electrical signal. When the transceiver passes this alternating electrical current to the antenna, electrons inside the atoms comprising the antenna begin to vibrate, and their vibration creates an oscillating electric field. This oscillating electric field in turn creates an oscillating magnetic field and as a result, you get electromagnetic waves.

These waves radiate from the antenna in all directions. All electromagnetic waves travel at the speed of light. In fact, light itself is an electromagnetic wave, but at a much higher frequency (this causes light to behave as both a wave and a particle). When the waves reach a receiving antenna, the process just described occurs in reverse — arriving electromagnetic waves excite the electrons in the antenna, creating an electrical current that the transceiver translates into data.

The frequency of an electromagnetic wave is determined by the *wavelength* or *wave cycle*. You determine the wavelength of a signal by measuring the distance between peaks in the wave (see Figure 8-10), in turn you determine the frequency by counting the number of *wave cycles* that pass a given point in a specified length of time. Science uses *hertz* (Hz) as the measurement of frequency, one Hz being one wave cycle passing a fixed point in one second (see Figure 8-11).

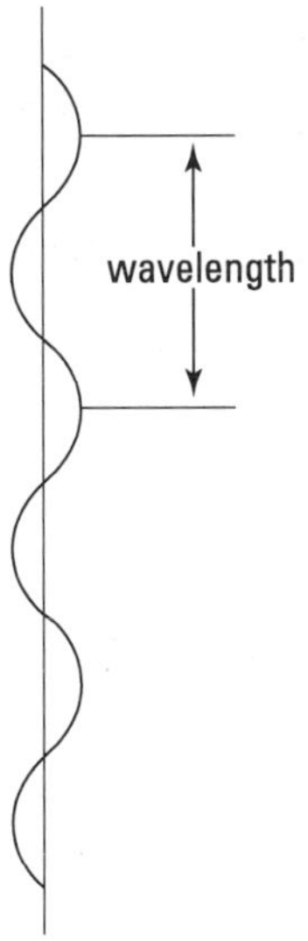

Figure 8-10: Wavelength

Long wavelength = low frequency   Short wavelength = high frequency

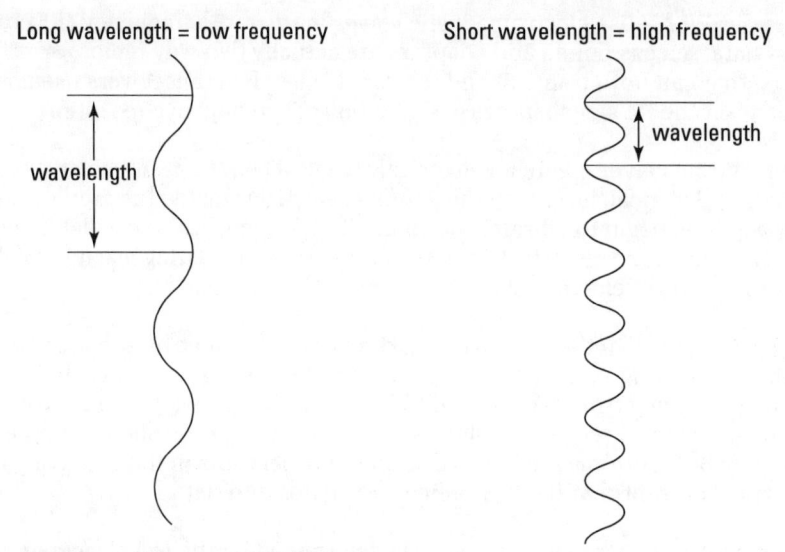

Figure 8-11: Wave cycle and frequency

Low-frequency signals have long wavelengths because it takes longer for each wave cycle to pass a given point. The signal from your WLAN hardware has a very short wavelength: 2.4 billion wave cycles per second for 802.11g and 802.11b (2.4 GHz), and 5 billion cycles per second for 802.11a (5 GHz).

An antenna can increase the strength and range of a signal. This is called *gain*, and it's measured in *decibels* (dB). Gain is the amount that the output power of any signal is amplified compared to the input power.

An antenna can't add power to a signal; it's just a piece of metal. What an antenna does to create gain is focus the signal. A good analogy is a lighthouse, which uses its lenses to focus the light it creates into a bright beam. Similarly, antennas create gain by focusing electromagnetic waves and creating a stronger radio signal.

Improper mounting

Improper mounting of an antenna can create interference and multipath problems. It's important that you follow the manufacturer's instructions carefully to avoid these issues. Some antennas will only operate correctly when mounted in a specific manner. For example, some must be mounted vertically instead of horizontally. Failing to mount the antenna correctly may reduce your Wi-Fi signal coverage (see Figure 8-12).

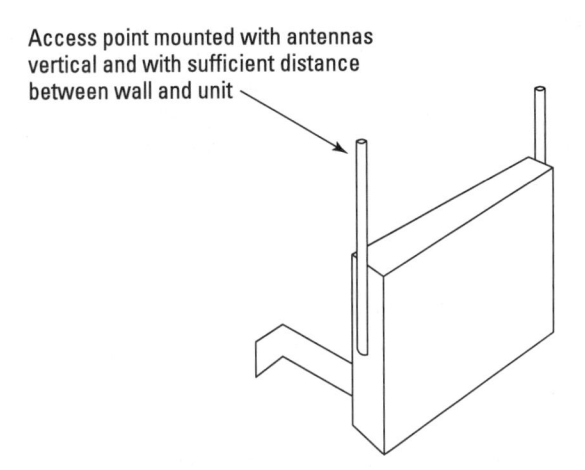

Access point mounted with antennas vertical and with sufficient distance between wall and unit

Figure 8-12: Proper mounting of an antenna

You must also follow instructions regarding distance from other antennas, walls, and roof. Mounting your antenna directly adjacent to another antenna may create problems from interference. Similarly, mounting an antenna or access point too closely to a wall, roof, or metal and concrete objects may create multipath interference (see Figure 8-13).

Mounting antennas too close to one another or too close to the roof surface can create multipath problems and other forms of interference

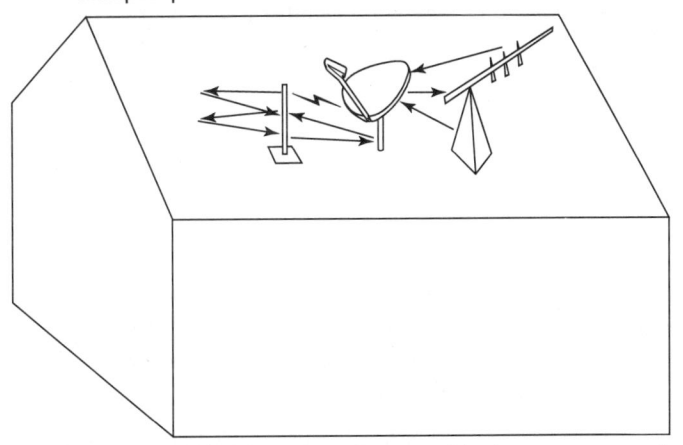

Figure 8-13: Multipath interference created by improper mounting

To avoid creating mutlipath problems due to improper mounting of an access point, be sure that the antennas have 6 to 8 inches of clearance from the walls or ceiling whenever you wall-mount an access point or omnidirectional antenna.

Note: An omnidirectional antenna radiates its RF signal in all directions simultaneously. Omnidirectional antennas are standard on most access points.

Grounding antennas

Whenever you install an external antenna, including a satellite or dish antenna, you must be sure it is properly grounded. If you fail to ground your system properly, a static charge can build up, creating interference and discharge that will damage your equipment.

The ideal method for properly grounding your antenna is through the house's electrical system. This is a job for a professional installer, partly because of compliance requirements with local building codes, but also because inexperience in this area can lead to severe injury and even death.

Caution: The previous warning is worth repeating: if you make a mistake while trying to ground your system, it could kill you. Don't do this yourself unless you're a professional installer and you know exactly what you're doing.

Grounding an antenna to a ground pole that has been driven into the ground is insufficient. The charge at the ground rod won't be equal to the charge at the equipment, and the static discharge that results will destroy your equipment. Grounding through a building's power supply will prevent this (see Figure 8-14)

Often, if a system has been without power, such as after a blackout, or if it's been shut down for a significant period, a static charge can build up and cause interference once you power the system up again. You can correct this by manually discharging the static charge in the antenna cable.

Caution: Make sure that you turn off power at both ends of the antenna cable in the event that there is powered equipment at both ends.

To do this, you must first turn off all of the power to the system. After you have verified that the power is off, remove the antenna cable from the back of your access point or satellite modem. While holding the cable, touch your thumb to the center pin of the cable (see Figure 8-15). This should eliminate the static in the line. Then you can reattach the cable and restore the power.

Proper grounding protects your equipment from lightning strikes as well as damage or interference from static charges. Many wireless service providers require that your installer ground the antenna or your warranty will be voided.

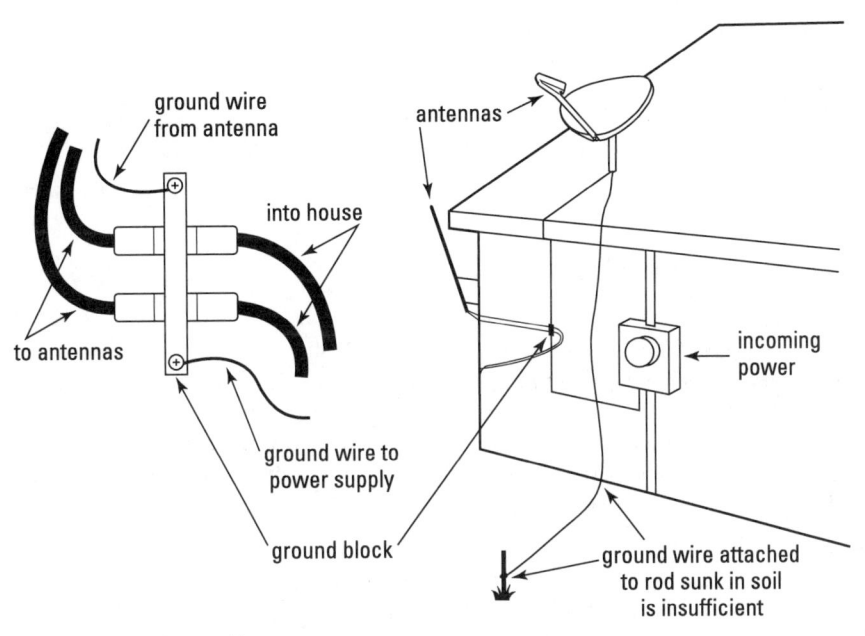

Figure 8-14: Grounding an antenna

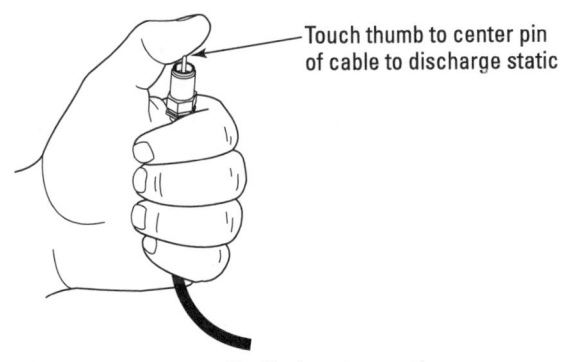

Figure 8-15: Manually discharging static

Delineating the Fresnel zone

If you have an outdoor line-of-site connection between two antennas, one important thing you have to consider is the *Fresnel zone*. Named for the French physicist Augustin-Jean Fresnel (pronounced fray-nel), the same man who designed the lenses first used in lighthouses, the Fresnel zone is the pattern of RF radiation between two antennas (see Figure 8-16). The Fresnel zone is shaped liked an elongated ellipse and must remain free of obstructions in order for a line-of-site connection to operate correctly.

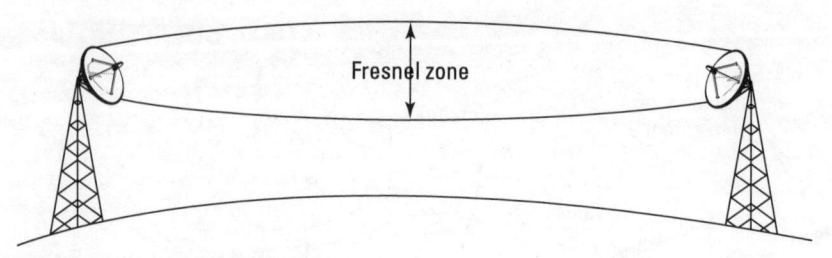

Figure 8-16: The Fresnel zone

Objects, including trees, that extend into the Fresnel zone between two antennas will cause increased signal *attenuation* or interference (see Figure 8-17). Generally, in order for a link to operate efficiently 60 to 80 percent of the Fresnel zone must remain free of obstructions. You can accomplish this by mounting your antenna high enough to allow the Fresnel zone to remain above objects in the signal path.

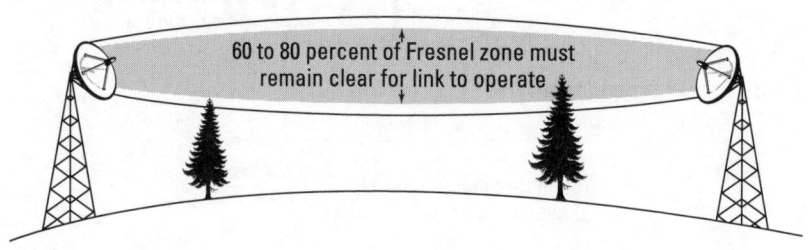

Figure 8-17: The Fresnel zone must remain free of obstructions.

The size of the Fresnel zone depends upon the distance between the antennas and the frequency of the signal. For 2.4 GHz Wi-Fi, it is roughly 14 feet for a link distance of one mile and 10 feet for a half-mile link. For short-distance links, allow at least 10 feet for the Fresnel Zone, which should provide you with plenty of wiggle room when planning the link.

Interference from trees

Trees are a possible source of interference in outdoor links. Thick vegetation can contribute to Wi-Fi signal attenuation. This is partly because trees, like many living things, are almost 80 percent water (the leaves at least) and water can slow a 2.4 GHz signal.

> **Note** When the 2.4 GHz RF radiation passes through water, the signal loses some of its power, which is transferred to the water as heat. This is why a microwave oven (also 2.4 GHz) can boil water.

Chapter 8 ✦ **Technical Pitfalls and Solutions** 145

When you plan an outdoor link between buildings, make sure that you have an unobstructed line of site and that trees, shrubs, or heavy vegetation will not obstruct the signal. If you do your planning during the winter, remember that while a link may work fine when trees are bare, come spring, you may be in for a surprise when the leaves return and the signal degrades (see Figure 8-18).

Trees can weaken Wi-Fi signal when foliage returns in spring

Figure 8-18: Interference from trees

Optimizing Equipment

There are ways to tweak the performance of some networking devices such as broadband and satellite modems, but the specifics are unique to each type of device, which makes it impractical to address them within the scope of this book. However, you can find tips specific to your modem and Internet service (DSL, cable, or dial-up) online.

Some Web sites that offer tips to improve the performance of your broadband and wireless devices are:

- www.broadbandreports.com
- www.speedguide.net
- www.tweak3d.net

There are also steps that you can take to improve the performance of your wireless devices and network clients, such as updating drivers and upgrading firmware. Checking for driver and firmware updates is one of the first things you should do when you get your new device home. Chances are that it's been sitting on a shelf for a while, and during that time the manufacturer has probably made improvements.

Drivers are software modules that enable your operating system to communicate and operate with your device. *Firmware* is the operating software for a hardware device, and it is located in programmable memory chips within the device. You can *flash*, or reprogram, these programmable chips (EPROM, ROM, PROM) using update software supplied by the manufacturer.

Manufacturers use firmware updates to extend the useful life of equipment, improve performance, and add new features. One of the best ways to ensure that you get the best performance from your gear is to diligently apply firmware and driver updates as they become available.

Summary

As I stated at the beginning of this chapter, there are a number of things that can affect the performance of your WLAN. In this chapter, I explained some of the more common ones and their resolution. These included:

- Multipath interference issues
- Wi-Fi channels overlapping
- Theft of service
- Configuration errors
- Dealing with RF interference inside and outside of your home
- Interoperability issues
- Mounting and grounding antennas
- Optimizing equipment

You can trace most Wi-Fi problems to one of these issues. Dealing with these problems during planning and installation will help you to avoid trying to identify and deal with technical problems after your WLAN is up and running.

✦ ✦ ✦

Wireless Privacy Concerns

CHAPTER 9

In This Chapter

Understanding why privacy matters

Threats to Wi-Fi privacy

Guarding against privacy threats online

Reading privacy policies carefully

Vulnerable non-Wi-Fi wireless technology

Most of us want to protect what privacy we have left. A wireless network can threaten your privacy if it isn't properly secured. Even then, there are threats both online and offline that can undermine your privacy and expose your personal information to theft. This chapter discusses privacy issues and how to protect your personal information both on wireless and wired networks.

Why Privacy Matters

Privacy rights can be a polarizing issue in our current political climate. If you're concerned about your privacy rights, you're liable to be labeled as paranoid. Unfortunately, many people see the issue as black and white, thinking, "I have nothing to hide; people so concerned about privacy are probably up to something no good."

The problem is that many people fail to see privacy as a security problem. In this context, all of us have something to hide and protect. It's our identity. Identity theft is quickly outpacing all other electronic crimes facilitated in part by eroding privacy protections and massive data collection by corporations. It isn't *Big Brother* you need to be concerned about, it's *Little Brother, Inc.*

Corporations have created huge databases of our shopping habits, financial data, age, health, hobbies, and more. Actually, they've been doing this for years; the new problem is data sharing (or selling) and data mining. Bits and pieces of information from hundreds of seemingly unrelated databases can combine to create accurate personal profiles.

Unbelievably, much of this information finds its way into public databases, either available for free on the Internet or by subscription. If you use the Internet a lot then you've probably left an even easier trail to follow. Everything you've posted or written on the net has been indexed and cached somewhere, just waiting for someone to retrieve it.

The amount of data accessible in public and private databases is staggering. In a few minutes, if you know where to look, you can gather enough information about the average adult U.S. citizen to assume their identity and commit fraud. If you're willing to pay for access to private databases, you can find out even more.

As a test, a client asked me to see what I could discover about him, strictly using the Internet, and not using any of the usual channels I would use in an investigation or background check. In one hour, I had a complete background, including the following:

- His birthdate
- His mother's maiden name
- His home address and phone numbers
- Five profiles, under aliases, on MSN and Yahoo
- His mortgage deeds and related records
- Partial health history
- Where he went to high school
- Cell phone number
- Five e-mail addresses
- Divorce and marriage records
- His Social Security number

When I began all I knew was his name and where he worked. All I used for research was the Google search engine and a few public databases. In fact, public databases are becoming so complete that many law enforcement agencies use them in investigations. One federal agent that I know swears that Google is the best thing to happen to law enforcement in years.

The insecurity of wireless networks and emerging technologies for tracking mobile wireless users are compounding existing privacy problems. In order to protect your privacy, you first need to understand these threats. The following sections present these issues.

Wi-Fi Privacy Threats

In addition to privacy threats due to the Wi-Fi security problems outlined in this book, there are new issues arising as Wi-Fi access expands and more people connect to the Internet through hotspots.

While you may have secured your WLAN at home, when you go to a hotspot you've left that protection behind. There is no guarantee that the operator of the hotspot has taken any steps to protect his network or that your network communications aren't being monitored.

There's nothing to stop a rogue hotspot operator from recording where you go on the Internet or even from intercepting your passwords and personal data. Even legitimate service providers can collect data on their customers' online activities and geographic locations. Being aware of this potential for abuse is the first step toward protecting your privacy.

Understanding location-based services

Location-based advertising is a new idea that exploits a wireless service provider's ability to pinpoint a user's location and use that information to deliver location-relevant advertising. Yes, that's right; I said *pinpoint* your location. All cellular phone companies have the ability to triangulate the position of any phone connected to their network. Most are expanding this ability because of new federal regulations requiring these companies to be able to locate a customer who places an emergency 911 call.

Pinpointing your location doesn't require that a company use the global positioning system (GPS); it simply monitors the signal strength among multiple cellular towers and calculates the time it takes for the phone's signal to reach each of them, or triangulating its location (see Figure 9-1). However, many new phones do feature GPS capability and carriers are sure to leverage this.

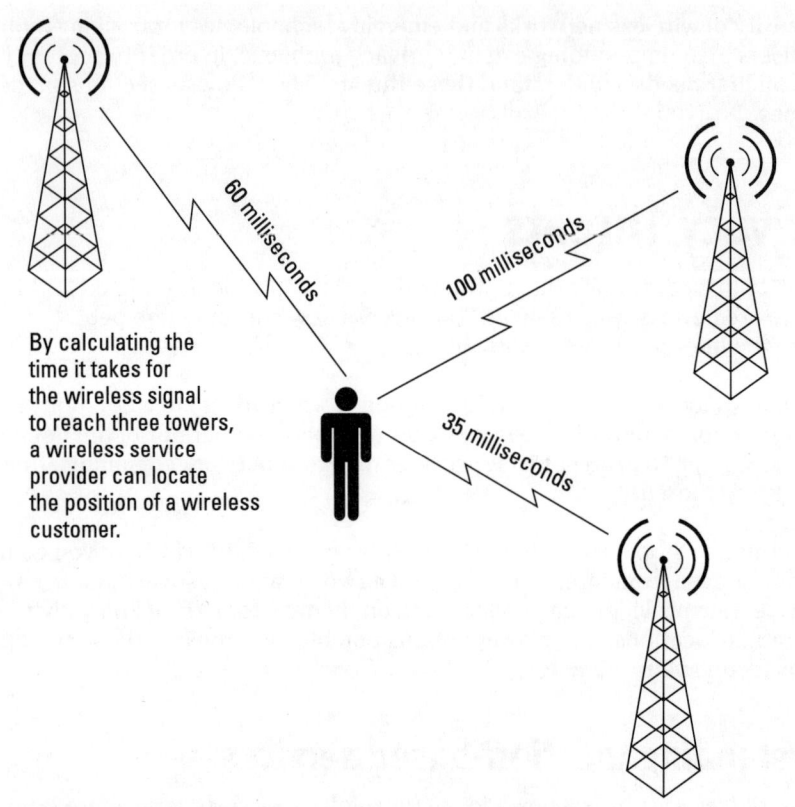

By calculating the time it takes for the wireless signal to reach three towers, a wireless service provider can locate the position of a wireless customer.

Figure 9-1: Triangulating a wireless customer's position

Once the carrier knows a user's location, it can deliver relevant advertising. For example, if I were in a downtown area and a nearby electronics store was having a sale, then my wireless could send an advertisement about the sale directly to my phone when I was near the store (see Figure 9-2).

This doesn't only apply to wireless phones; advertisers are targeting Wi-Fi devices as well. In April 2004, a company called Quarterscope announced it had developed a *Wi-Fi positioning system* (WPS) that it planned to have in service by the end of the year.

Advertising isn't the only application of this technology. Quarterscope is planning a number of applications, such as wireless city guides, that will take advantage of this positioning system. However, it is also going to market a people-tracking service that presumably parents can use to monitor their kids' movements.

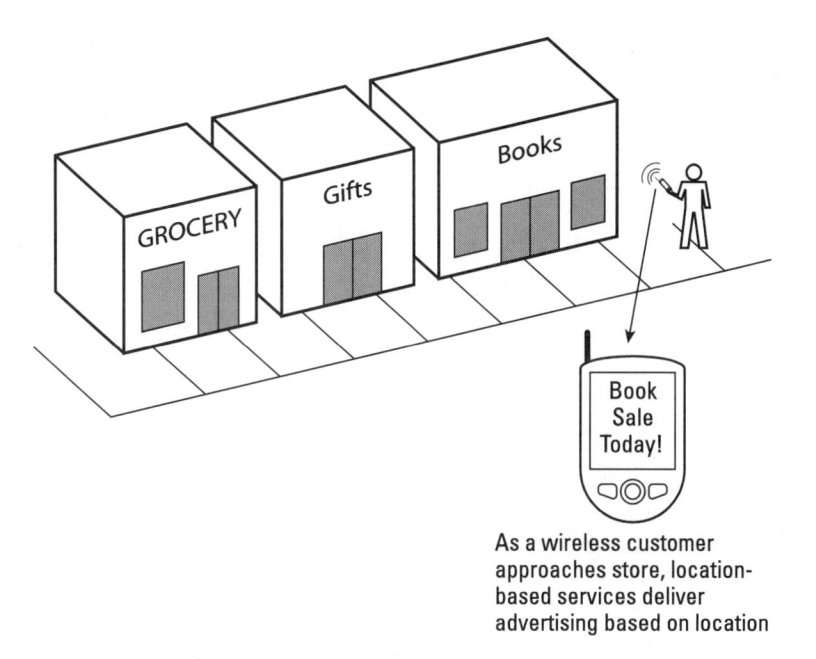

Figure 9-2: Location-based ad delivery

As ISPs and advertisers begin to track your day-to-day movements and store this data, you have to be aware that this information is likely to find its way onto the Internet or into commercially accessible databases. The potential for abuse of this information is considerable; it adds a completely new dimension to cyber-stalking.

The number of online threats to your privacy continues to grow. Whether you access the Internet via Wi-Fi or a wired connection, you'll encounter these problems and will have to take steps to protect yourself.

Exposing spyware

Spyware monitors your Web surfing and may record everything you type, including passwords and credit card numbers. Spyware can end up on your computer in different ways. Some may arrive as part of a worm or virus, while other spyware utilities may be carried by a Trojan program and install themselves whenever the program executes.

> **Cross-Reference**
> Refer to Chapter 7 for more information about worms, viruses, and Trojans.

Spyware might be installed in a drive-by download or a person with access to your computer may install a keylogger or similar application to steal your passwords and record your conversations. The best way to prevent spyware from invading your computer is to do the following:

- Install antivirus software and keep it updated
- Install antispyware software and keep it updated
- Install a personal firewall application
- Don't open or download suspicious e-mail attachments or software
- Monitor who has access to your computer

These simple steps will help prevent spyware from violating your privacy and security.

If you believe your computer has spyware on it or if you'd just like to check it to be sure, you can download antispyware software that will do the job. The next section lists some useful software for blocking and removing spyware.

Avoiding adware and drive-by downloads

The drive-by download is a new tactic of overaggressive Internet advertisers and, in some cases, crackers. The name drive-by comes from the fact that you'll be surfing the Web, and while visiting a Web page, the site secretly downloads an application to your computer.

This attack uses misleading tactics to get you to authorize a download, such as a pop-up menu that says "Click here to enter a contest," or exploits a security flaw in your browser to install itself without you even knowing.

The application downloaded to your computer is usually some sort of *adware*, which is software that's used to monitor your activities and deliver ads to your computer. Occasionally, you may also find your computer has been infected with spyware in a drive-by download.

Adware can also be in the form of cookies that collect personal data and track your visits to different Web sites. These are sometimes called *Internet transponders* because they monitor where you're visiting on the Internet and report it back to a central service.

Once you're the victim of a drive-by download, you'll be inundated with pop-up advertising, and advertisers may even track your movements and send you targeted ads based on your Web searches and pages you visit.

Unfortunately, adware and even spyware can be hard to avoid. One visit to the wrong Web site and you're infected. Antivirus software doesn't defend against many of these applications, and a firewall doesn't provide any protection at all, because the adware is downloaded via your Web browser, often with your unwitting consent.

Caution Adware and spyware can also be installed on your machine as part of another application. Many free programs are advertising-supported and have ad software built in. Others may track you or collect information about your surfing habits. Most software that includes adware discloses this in the End User Agreement prior to installation. Make sure you read this carefully and decide if you want to proceed.

Some steps you can take to protect yourself (and your sanity) from adware and spyware are:

- Keep your operating system and browser patched and updated
- Keep your antivirus software updated
- Be careful which sites you visit and what you agree to
- Be careful which free applications you install on your machine
- Install adware blocking/removing software on your machine, and keep it updated

Spy-blocking and ad-blocking software is useful for protecting your machine from infection and cleaning it if it is infected. There are a number of good programs in this category, some are freeware and others are shareware or commercial products. Some of the popular antispy and antiadware programs are:

- **Ad-aware.** Developed by Lavasoft and a free version is available at `www.lavasoftusa.com`
- **Spyblocs.** Developed by Eblocs and a limited free version is available at `www.eblocs.com`
- **Spybot search and destroy.** From Safer Networking Ltd. and a free download is available at `www.Spybot.info`

Several sites provide information and links to download antiadware and antispy programs. These include:

- www.securityconfig.com/software/desktopsecurity/desktopsecurity.htm
- www.spychecker.com
- www.spywareinfo.com

Tossing your cookies for greater privacy

Cookies are passive files, not executable applications. They are nothing more than text files that contain information. Web sites can create cookies on your machine in order to facilitate services or to track your surfing habits. Tracking cookies work by using a network of participating sites to track your movement.

When you visit a Web site, it checks for the tracking cookie and notes your visit. As you surf the Web, participating Web sites read the cookie and report your visits back to a central server, which then builds a profile of your habits.

In some cases you're identified by nothing more than a random serial number, but there have been cases where sites have collected more personally identifiable information, including e-mail addresses, ISPs, and even names.

Fortunately, you can control cookies, either with antiadware programs or through adjusting your browser settings. Antiadware programs look for tracking cookies and cookies that contain personable identifiable information and delete or block them. In Internet Explorer you can change your privacy settings. Follow these steps:

STEPS: Adjusting your privacy settings in Internet Explorer

1. **Launch Internet Explorer.**
2. **Click Tools in the menu bar and select Internet Options from the menu.** The Internet Options dialog box appears (see Figure 9-3).
3. **Click the Privacy tab to view the Privacy settings options, as shown in Figure 9-4.**
4. **Move the slider to adjust your privacy settings.** As you move it up or down you'll see definitions of the various levels of privacy to the right of the slider.

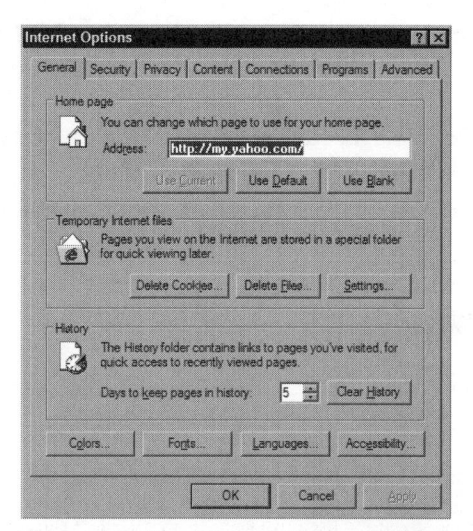

Figure 9-3: The Internet Options dialog box

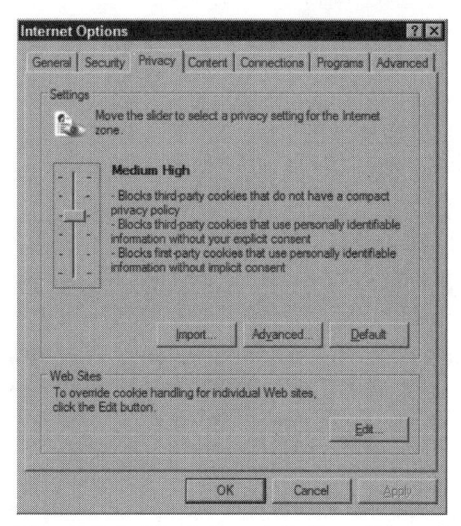

Figure 9-4: The Internet Options Privacy tab

5. **To override the privacy settings and allow or block specific sites from setting cookies, click Edit.** The Per Site Privacy Actions dialog box appears (see Figure 9-5).

6. **Enter a Web site you want to manage; click Block or Allow.** When you're finished click OK. The Per Site Privacy Actions dialog box closes.

7. **Click OK again.** The Internet Options dialog box closes and your settings will take effect.

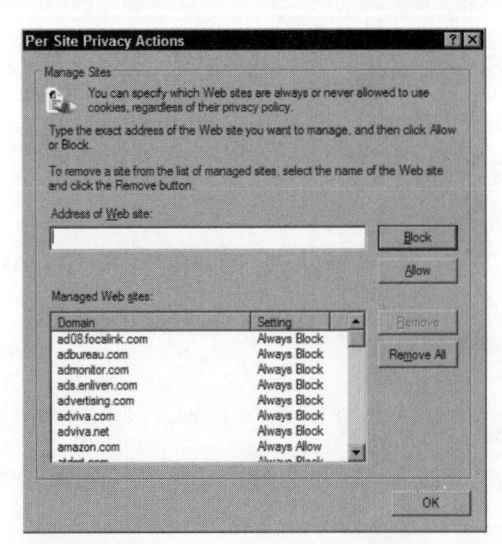

Figure 9-5: The Per Site Privacy Actions dialog box

By adjusting you browser's privacy settings, you can block sites attempting to create cookies that contain personal information. Alternately, you can manually delete cookie files or use ad-blocking software to augment Internet Explorer's ability to block undesirable cookies.

Demystifying Privacy Policies

A company or Web site's privacy policy details how that organization collects data and what it does with that data afterward. Specifically, a privacy policy should tell you the following:

- What type of information the organization collects
- What it does with that information
- With whom it shares the information
- What steps it takes to protect your privacy

The type of information the organization collects is important. Can it or its partners use the information to identify you? Does it provide enough information to enable the organization to contact you directly? It's important that you pay close attention to who is collecting personal data and what is done with it.

While you may not care if a Web site uses your e-mail address to send you commercial e-mail informing you of new products and offers, it may become an issue if the site gives or sells that information to other sites. Before you know it, your e-mail address could be on every spam list in the country.

Most Web sites, especially commercial sites, have online privacy policies that you can read and review. Often, you must do this as part of your registration. Don't just click Yes and move on, however. Take the time to actually read the policy and be sure that you aren't giving away too much of your right to privacy.

Some Web sites may use a *Platform for Privacy Preferences* (P3P) policy. P3P is a technical standard that describes a standard format for privacy policies that Web browsers can retrieve automatically and interpret based on your browser's privacy settings. Based on your preferences and the company's P3P policy, your browser can determine whether to allow the site to collect information or place a cookie on your machine.

Another group of policies to which you should pay close attention are those that belong to wireless service providers, specifically companies that operate Wi-Fi hotspots. Some questions you should ask are:

- Does the company collect geographic data or track your movements?
- With whom, and under what circumstances, is this information shared?
- How long is tracking data stored?

Companies can only collect data that you allow them to, either by you supplying it or allowing them to collect it through observation. By being aware of who's collecting data about you and why, you'll be better prepared to protect your identity and limit your exposure to threats like identity theft.

Other Vulnerable Wireless Technology

Clients who use 802.11 wireless networks aren't the only wireless Internet users subject to eavesdropping and tampering. The number of available consumer wireless products is growing amazingly fast. As more products offer wireless connectivity, take care that you aren't trading privacy for convenience.

Throughout this book, I've illustrated the security and privacy issues inherent with wireless computer networks. In fact, I've spent several chapters addressing the problems with Wi-Fi encryption, the number of ways crackers can attack your network, as well as technical problems that create security and safety issues for users.

Despite this, and even though there are plenty of reasons to be concerned about Wi-Fi networks, wireless computer networks do have some protections built in. However, the majority of other consumer wireless devices don't have any sort of built-in security. Most of the time, security is an afterthought or simply ignored during the design phase.

Of all the available wireless products, three groups pose the biggest threat to your privacy and, in some cases, security. These are:

- ✦ Wireless home controllers, including X10 devices
- ✦ Cordless phones, including 2.4 GHz models
- ✦ Wireless video cameras

Intercepting X10 device signals

X10 Inc. introduced the X10 standard for home controllers. Most X10 devices use the electrical wiring in a house as their network medium. Once they're plugged into an outlet they can communicate with other X10 devices, sending and receiving instructions.

Many newer X10 devices, including remotes and cameras, are now wireless and usually use the same 2.4 GHz band as 802.11 devices. The majority of them have no security features whatsoever. Anyone can receive, transmit, or eavesdrop on the signals as long as they have some inexpensive equipment.

Usually, all an intruder or a snoop needs is a compatible X10 device, either a controller or a receiver. Let's say that you've used X10 devices to set up an automated home control system. For ease of use, you've installed X10 wireless adapters for your controllers so that you can control your lights, doors, and appliances from anywhere in the house with your wireless X10 remote (see Figure 9-6).

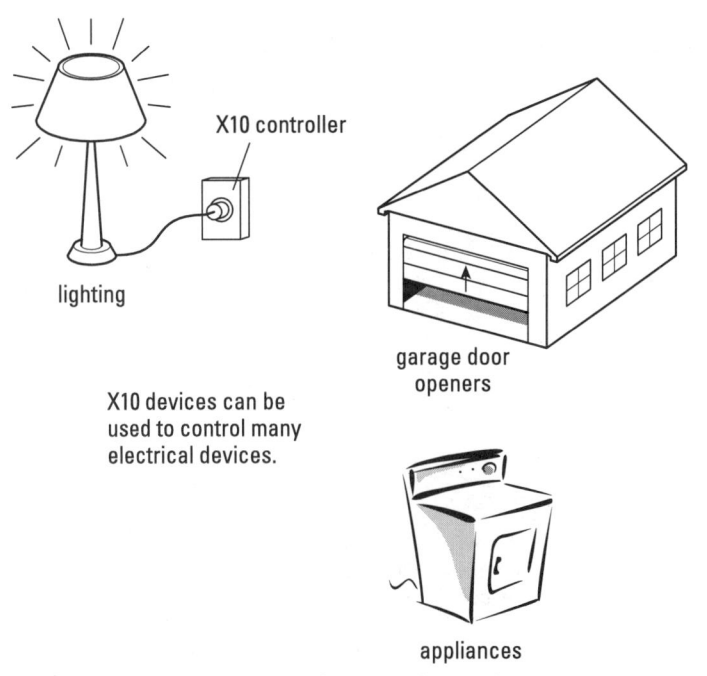

Figure 9-6: Automating your home with X10 devices

If I know this, and I want to get into your house or just tamper with your automation, like repeatedly turning your lights off and on, all I have to do is purchase an X10-compatible wireless controller and figure out your house and device codes. Once I have those programmed, I can take control. If you have X10 locks on your doors, I can open them. If you have an X10 controller for your garage door opener, I can open that, too.

Figuring out the codes is trivial; each X10 device can be set to one of 16 house codes and 16 unit codes (see Figure 9-7). That means I would only have to try each of the 256 possible combinations until I get in. That would probably take less than 10 minutes.

X10 also offers wireless video cameras; at some point you've probably seen one of the millions of Internet ads for these devices. For a couple of years you could hardly launch your Web browser without a series of X10 pop-up ads confronting you. Most of these ads are gone now, but the cameras remain. By some estimates, X10 sold a couple million of these in the U.S.

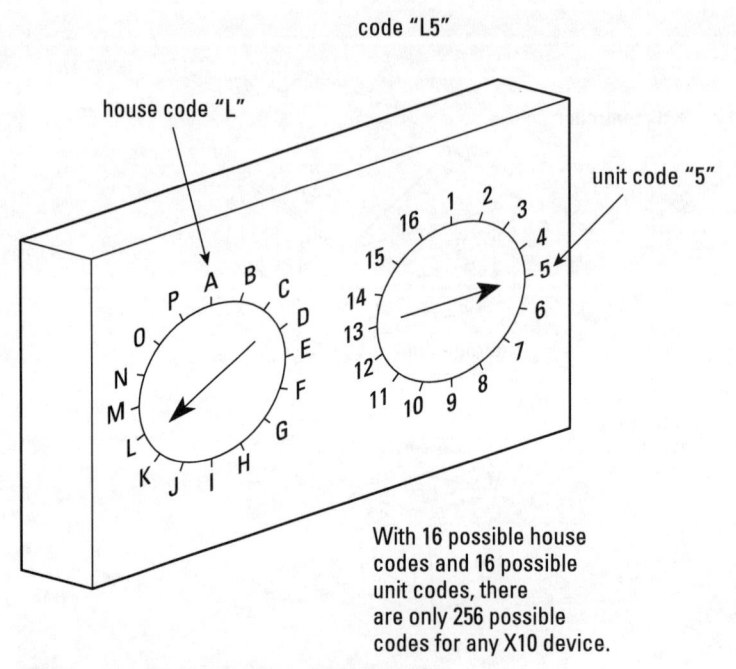

Figure 9-7: Setting an X10 device's code

With 16 possible house codes and 16 possible unit codes, there are only 256 possible codes for any X10 device.

Most of these cameras operate in the 2.4 GHz band, like Wi-Fi gear. The majority of these cameras have only three channels and absolutely no encryption or security at all. To intercept the signals from these cameras, all you need is an X10 receiver plugged into a TV (see Figure 9-8). Because there are only three channels from which to choose, it's easy to find the signal, even if you have to change channels manually.

> **Cross-Reference**
> In Chapter 6 I discuss warspying, the term used to refer to intercepting wireless video signals.

X10 devices are useful and home automation can be fun. You just need to be aware of the related risks and take a few steps to protect your home and your privacy. If you are using X10 or X10-compatible equipment marketed by companies like Radio Shack (Plug-N-Power), Sears, Stanley, and General Electric, consider the following precautions:

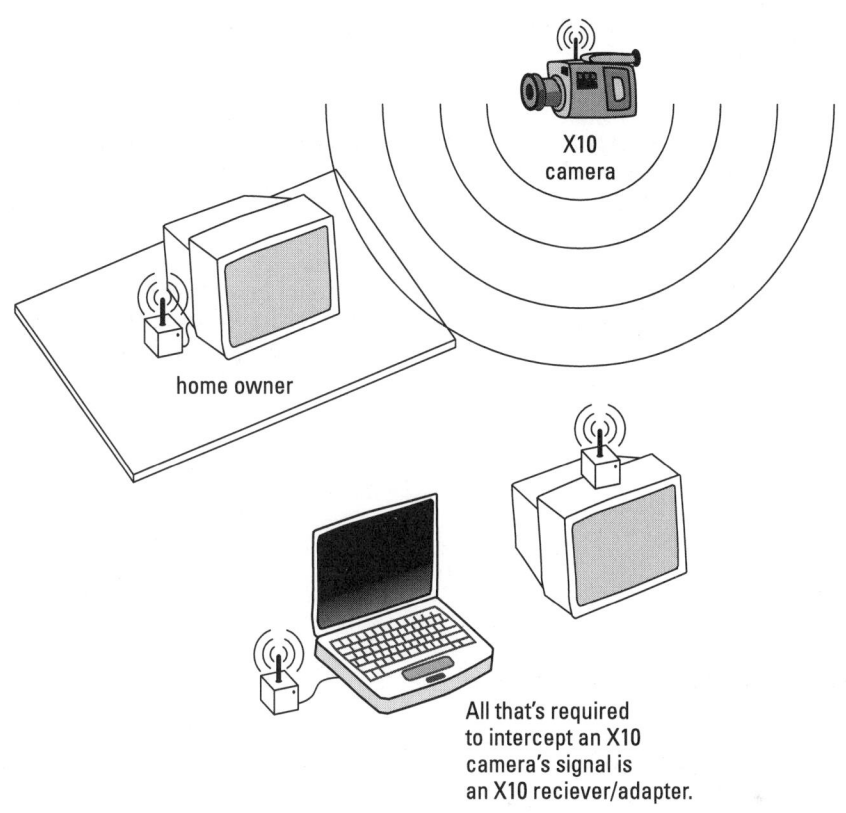

Figure 9-8: Intercepting an X10 video signal

- If possible, avoid using wireless controllers.
- Don't use X10 devices to control door locks, garage door, or other entrances to your home, even if they aren't wireless they're easy to tamper with.
- Avoid X10 cameras. If you do use one, don't point it at anything you don't want the neighbors to see (shower-cam anyone?).

Peeping in on Wi-Fi video cameras

In addition to the X10 cameras, there are many different brands of wireless 2.4 GHz cameras available. I distinguish these from X10 and other wireless cameras in that they're 802.11 compatible. Therefore, these are Wi-Fi cameras, where the remainder are merely 2.4 GHz cameras.

These cameras are usually network ready and compatible with one or more 802.11 standards (802.11a, 802.11b, or 802.11g). Each camera has its own IP address, and usually all that's required to view the image is a compatible Web browser. Some have optional password protection, and even when this is an option, it's not often used.

All you need to know to intercept the video feed from one of these cameras is its IP address. Any cracker can use a network packet sniffer to discover this. Then all the cracker has to do is input the IP address into a Web browser and watch the video (see Figure 9-9).

Cross-Reference To read more about sniffing refer to Chapter 4.

Wi-Fi camera
IP address 192.168.0.175

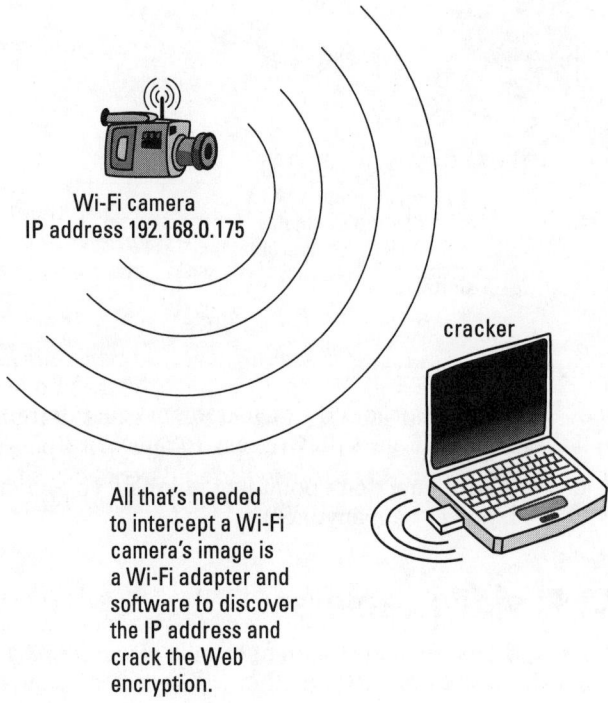

cracker

All that's needed to intercept a Wi-Fi camera's image is a Wi-Fi adapter and software to discover the IP address and crack the Web encryption.

Figure 9-9: Intercepting a Wi-Fi video signal

Some cameras offer WEP encryption as an option, but this offers limited security because WEP is easily broken. Because Wi-Fi cameras generate a lot of traffic, they can actually make it easier for a cracker to collect enough data packets to crack the encryption in a shorter amount of time.

Cross-Reference In Chapter 10, I discuss WEP encryption and how it's defeated in detail.

Wi-Fi cameras are somewhat more secure than X10 and other non-802.11 cameras. They're harder for casual wardrivers or warspyers to detect and not as simple to view. If you decide to use one, you can take steps to protect your privacy. These include:

- Don't point it at anything you wouldn't want the whole world to see. Be aware, some of these cameras can be controlled from a Web browser and have remote pan-and-tilt features. Therefore, a cracker could change the viewing area.
- If it's available, enable WEP encryption.
- If the camera offers any other security features such as SSID or MAC filtering, use them.

Eavesdropping on cordless phones

Cordless phones are a fixture in a majority of U.S. homes. Most of us use them without a second thought and assume that they afford us the same degree of privacy as ordinary phones with a cord attached. Unfortunately, they don't, and many people aren't aware of this.

Like other wireless devices, cordless phones and base stations are *transceivers* (transmitter-receivers) or two-way radios. When you use your cordless phone, the signal just doesn't travel between your handset and its base station; it radiates in all directions until it fades, bounces, or some obstruction blocks it. It's possible to intercept the signal from some cordless phones using a radio frequency scanner. Older phones, operating at 900 MHz and below, are particularly susceptible to eavesdropping.

Note In the United States, eavesdropping on any telephone conversation is illegal. Even police agencies have strict guidelines that they must adhere to regarding wiretaps.

Note Most cordless phones that operate at 900 MHz and below are still analog, use no encryption, and are not secure.

When the Federal Communications Commission (FCC) opened the 2.4 GHz and 5.0 GHz frequency bands in 1998, manufacturers began producing 2.4 GHz phones. Because of the higher frequencies, fewer scanners are available that can receive the signal. Note I said fewer, not none. Phones that operate in these frequencies are digital and often have some sort of proprietary encryption.

> **Cross-Reference** In Chapter 10, I discuss encryption types and their uses.

In an attempt to achieve security, the U.S. government outlawed the sale of scanners that can receive these high-frequency signals. Legislation is an ineffective stopgap for poorly designed security, however, as there are still plenty of these scanners available.

In 1995, cordless phone manufacturers began using Digital Spread Spectrum (DSS) technology to secure their phones. DSS spreads the signal from a conversation across several radio channels, making it much harder to intercept, let alone decrypt.

Being aware of the security concerns surrounding wireless phones is the first step toward greater privacy. Some other steps you can take to protect yourself are:

- If you have an analog cordless phone, replace it with a new digital model
- Consider one of the newer 2.4 GHz, or 5.8 GHz digital phones (but be aware that they may conflict with the signal from your WLAN)
- Be sure that your phone has some sort of encryption and has DSS technology

These steps alone should make your conversations reasonably secure from anybody other than a police or government agency. These officials have resources at their disposal that will allow them to listen to your calls if they really want to.

If you're concerned about this, you live in a country with an oppressive government, or you work overseas and have to protect your company's intellectual property, then you should also consider taking the following steps:

- Avoid using a cordless phone. Overseas, some governments don't allow encryption in consumer devices or force manufacturers to design in backdoors that allow them to eavesdrop.
- Consider using an encrypted landline phone. Many companies and government facilities have these available. This is much stronger encryption than exists in cordless phones.

- Pay attention to your surroundings and note who could be listening. Not all eavesdropping is electronic.
- When discussing sensitive information, be sure that the party on the other end of the conversation isn't using an insecure phone, such as an analog cordless.

Be aware that even if you're using a cordless phone with encryption and DSS, the security exists only between your handset and the base station. Once the call reaches the phone lines, it's no longer secure, and anyone with the skills and resources can tap the signal (legally or illegally). This is true even if the second party in the conversation has a digital cordless phone.

Truly secure phones aren't cordless and encrypt the signal from one end to the other, similar to what a virtual private network (VPN) does for network communications.

Summary

In this chapter, I discussed privacy issues and protecting your personal information on both wired and wireless networks. The topics I covered include:

- Why you should be concerned about your privacy
- Threats to Wi-Fi privacy, including location-based services
- Privacy threats from adware and spyware
- Importance of understanding privacy policies
- Vulnerable non-Wi-Fi devices, including X10 devices, cordless phones, and wireless cameras

As with any technology, as long as we are vigilant and stay informed of the related threats to our privacy, we can take the proper steps to protect our personal information from fraud and abuse.

✦ ✦ ✦

CHAPTER 10

Encryption and Wi-Fi

In This Chapter

Explaining encryption

Introducing modern encryption techniques

Examining Wired Equivalent Privacy (WEP)

Introducing Wi-Fi Protected Access (WPA)

Using Pretty Good Privacy (PGP)

Encryption is one of the best tools available for securing your personal information and data. There are many different types of encryption; all of them have varying degrees of effectiveness. While using any encryption is better than using none at all, it's important that you understand the application and level of effectiveness of each type of encryption system. Failure to understand encryption and its proper use may lead to your belief that your Wi-Fi network is secure when it really isn't. This chapter takes encryption out of the windowless rooms and domain of the spymasters, and explains it in plain English so that you can make the best use of it.

Introducing Encryption

Encryption is the process of transforming data into unreadable code and then restoring it to its original readable form when it reaches its destination. Encryption is used to protect data from prying eyes, to authenticate users on a network, and as a form of access control to block unauthorized persons from accessing data. You can also use encryption to verify the integrity of data and ensure that it hasn't been tampered with or damaged in transit.

Before getting started, it's important that I define a few of the terms that are used in this chapter and that you're likely to encounter whenever you read about encryption. Understanding these terms is useful, because an increasing number of computing products include some sort of encryption, and you're likely to encounter these terms in product reviews, articles, and user guides.

The crypts

Many of the terms share the same root, *crypt,* which comes from the Greek word *Kryptos* meaning obscure, hidden, secret, and mysterious — a perfect fit for the science of obscuring and hiding messages.

Cryptography is the science of converting data into secret code. *Cryptographers* are the scientists and professionals that study cryptography and develop encryption systems, also called *cryptosystems*. *Cryptanalysis* is the science, or craft, of deciphering secret code and defeating encryption systems. *Cryptanalysts* are persons practicing cryptanalysis.

Are you still with me? Good, let's continue. A scientist usually engages in both disciplines (cryptology and cryptanalysis) while studying encryption. Part of developing a secure cryptosystem is the ability to apply cryptanalysis to discover and correct its weaknesses. Therefore, a cryptographer is usually also a cryptanalyst, and vice versa, although individuals may specialize in one discipline over the other.

Cryptology is a related science that deals with the mathematics underpinning cryptography and cryptanalysis. A *cryptologist* is a mathematician whose primary area of study is cryptology. Cryptologists are interested in all fields of mathematics, as advances in any area can enhance or defeat cryptosystems.

When you use an encryption system to encode data, you are *encrypting* the original data. When you decode, or return the data to its original, unscrambled form, you're *decrypting* it. Encryption describes the process of encrypting and decrypting data.

Cipher

The word cipher comes from the Hebrew word *saphar*, meaning to count or number. For our purposes, the definition of a *cipher* (also *cypher*) is the mathematical algorithm applied to data to *encipher* or encrypt it, creating secret code. Another related term, *decipher*, refers to decoding or decrypting encrypted data to return it to its original state. Encipher, encrypt, decipher, and decrypt are all used interchangeably.

Plaintext

Plaintext is the original data prior to encryption. Anyone with the appropriate application can read plaintext data. For example, anyone with Microsoft Word can open and read a plaintext MS Word document, because it is not protected by encryption.

> **Note** A password is not encryption and does not encrypt a document. Assigning a password to a Word document is just a form of *access control*, and a weak one at that. The document remains plaintext, and anyone who guesses or bypasses the password can read it. Recovering the password on a Word file is trivial, and there are many programs available on the Internet that facilitate doing so.

Occasionally you will read about flaws in systems that allow passwords and other sensitive data to be sent as plaintext or *in the clear*. This means that data traveling over a network connection is readable by anyone with the appropriate software. Needless to say, this is a bad thing, and any time you conduct business over a network connection or shop online, the connection should be secure and protected by encryption. Usually, your Web browser will provide some sort of indication that a connection is secure, either by notifying you or by indicating a secure connection with a padlock icon (see Figure 10-1).

Figure 10-1: Internet Explorer displays a padlock icon in the lower menu bar to indicate a secure connection.

Ciphertext

Ciphertext (also *cyphertext*) is encrypted plaintext. Using a cipher or cryptosystem, the original data is converted into unreadable code. Before it can be read again, ciphertext must be deciphered, or decrypted back into plaintext, and that requires the use of an *encryption key*.

Encryption key

An *encryption key* is an alphanumeric sequence that, as part of a cryptosystem, is used to encrypt or decrypt data (see Figure 10-2). In early cryptosystems, a key could have been something as simple as a substitution table (see Figure 10-3). In modern cryptosystems, the key is part of the mathematical equation that initiates the encryption or decryption process.

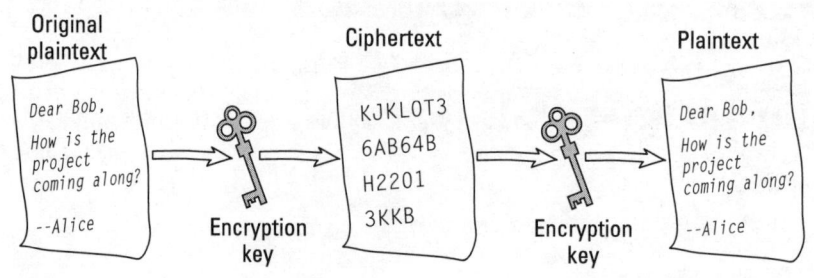

Figure 10-2: Encrypting and decrypting data with an encryption key

A = F	N = O
B = Q	P = R
C = X	Q = B
D = L	R = P
E = J	S = K
F = A	T = L
G = I	U = Z

Figure 10-3: A simple substitution table

A very brief history of encryption

The history of encryption and its related sciences, cryptography and cryptology, is fascinating. Governments and individuals have been using encryption to protect sensitive information for thousands of years. Many of the first encryption methods were developed by military forces to protect plans and messages. During this time, encryption has evolved from simple forms of letter substitution and transposition to mathematical encryption algorithms.

In ancient Greece, the Spartans developed a type of encryption that utilized a stick and belt. They called their method a *scytale* (pronounced skee-ta-lee). The sender of the message would wrap the belt around the stick and write the message on the belt. The messenger would then wear or conceal the belt while en route to the recipient. The recipient of the message would then wrap the belt around a similar stick to decipher the message, as shown in Figure 10-4.

The scytale was a form of *transposition cipher* and effectively transposed each letter a number of spaces, depending on the diameter of the stick. The stick acted as the *encryption key*. If an enemy intercepted the belt but used the wrong-sized stick, the message would be gibberish. The scytale was also a type of *steganography*, or hidden writing. It served to hide the message as well as encode it. Unfortunately, the stick was not an effective encryption key as anyone could try stick after stick until discovering the correct size.

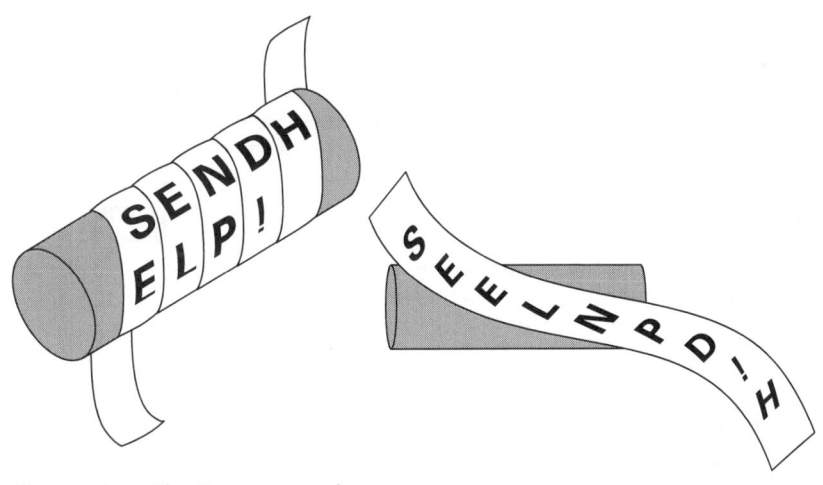

Figure 10-4: The Spartan scytale

The Romans used encryption extensively to protect military dispatches from the prying eyes of their enemies. Julius Caesar developed his own encryption method based on a letter transposition cipher. In a transposition cipher, every letter is transposed or rotated a certain number of letters. Caesar transposed letters in his cipher by three letters. In English, this would mean that the letter *D* would replace the letter *A*, and so on.

> **Note** People still use transposition ciphers. ROT13 encryption is a simple transposition cipher used to obscure Usenet news postings. ROT13 transposes every letter alphabetically by 13 spaces. ROT13 provides no real security as anyone can sit down and figure out the message by rotating the letter by 13 spaces again, returning them to the original text. However, it does confuse search engines and prevents indexing of messages.

For a long time, this was an effective means of encoding a message until people discovered a method of cracking this type of cipher called *frequency analysis*. In any written language, certain letters or sounds appear more often than others do. For example, vowels usually appear more often than consonants. In English, the most frequently appearing letter is *E*.

Ciphers that transpose or substitute letters may change the text of a message, but they don't alter the frequency of the letters or underlying patterns. If you substitute *K* for *E*, *K* will appear at the frequency normally attributed to *E*, and any frequency analyses will identify the letter *K* as the likely substitution for *E*.

The effectiveness of frequency analyses illustrates one avenue of attack against any encryption system: the identification of *patterns*. Whether the cryptosystem utilizes transposition, substitution, or modern mathematical techniques, cryptanalysts seek to identify patterns in ciphertext that they can use to infer or discover an encryption key.

In World War II, the German military developed a machine it called *enigma*. Enigma was a mechanical cipher that used rotating gears to substitute letters in plaintext to create ciphertext. Enigma's mechanism created sufficient entropy to thwart allied attempts at cryptanalysis. In fact, until the allies recovered an enigma machine from a captured German U-boat, the German's military communications were indecipherable.

Secret keys

A common problem shared by all early encryption systems (and some modern ones) is the distribution of the encryption key. All parties had to have a copy of the same key in order to communicate securely. Early cryptosystems and some modern ones as well relied heavily on secret keys (also called symmetric keys). If the keys were to fall into the hands of an enemy or foreign agent, all further communications would be compromised. Modern encryption techniques, utilizing computers and advanced mathematics, have solved some of the problems related to key distribution.

Understanding Modern Encryption Techniques

Modern cryptosystems use advanced mathematical algorithms to encode data, rather than simple transposition or mechanical encryption. Because computers can perform complex calculation of large numbers almost instantaneously, cryptographers have been able to develop incredibly strong encryption.

Encryption performs, or enables, two vitally important functions on computer networks: authentication and securing data. Some networking systems implement encryption to authenticate data packets and ensure that they originated from an authorized host. Other systems, including *Virtual Private Networks* (VPNs), go a step further and encrypt the contents of each data packet to prevent unauthorized persons from reading it.

Regardless of the particular cryptosystem, most employ some form of symmetric (secret) key or asymmetric (public key) cryptography.

Failings of secret key cryptography

As I mentioned earlier, secret key encryption systems were plagued by a common problem: distributing the key to everyone who needed it without revealing it to adversaries. This is the sort of stuff spy movies are based on — agent X must get the codes through to agent Y, or all will be lost.

In reality, distribution of secret keys actually was, and sometimes still is, a huge problem for intelligence agencies and military forces. As more business and organizations began to conduct business electronically through telecommunications or computer networks, the problem became an even bigger issue.

If your business relies on information remaining secure or being authenticated, such as in financial transactions, then it needs to be encrypted. How then, do you get the proper key to every node on your network? If you simply transmit the key in-the-clear, anyone who has access to your network could intercept it and use it to decipher your transactions.

The answer is to encrypt the secret key and then transmit it. Of course, if you encrypt the key, then you need to transmit another key for deciphering the first, and so on. It seems like a paradox; you must encrypt a key to protect a key, but then that key must be encrypted to protect it, and so on. Another encryption technology, called public key cryptography solved this problem.

Understanding public key cryptography

Public Key Cryptography (PKC), also called *asymmetric key* cryptography, was invented by Martin Hellman, Whitfield Diffie, and Ralph Merkle in 1976. Public key cryptography solved some of the problems associated with secret key systems. Chief among these problems was key distribution. You can use PKC to encrypt a secret key and distribute it to the appropriate persons.

In a public key system, encryption and decryption require two different keys, a public key to encrypt the plaintext and a private key to decrypt it. A *public key* is just that, public. Everyone can know your public key, and it will not compromise your encrypted communication. Only you have access to your private key, as it can decrypt all messages encrypted with your public key.

For example using a public key cryptosystem, Alice wants to send Bob a secure message. Alice uses Bob's public key to encrypt the message to Bob; she can then e-mail or transmit the ciphertext that will remain secure against prying eyes. When he receives the message, Bob uses his private key to decrypt the ciphertext into readable plaintext (see Figure 10-5).

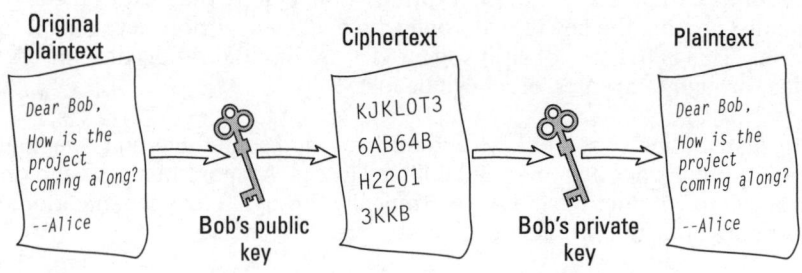

Figure 10-5: Using a public key system to encrypt and decrypt a message

Even though the whole world knows Bob's public key, it can't be used to decrypt his personal messages, only to encrypt them. It's also extremely unlikely that anyone will be able to use his public key to determine his private key because of the extremely difficult mathematical process involved in creating each key.

> **Note** Notice that I said it would be "extremely unlikely," but not impossible to discover a private key from a known public key. It depends on the particular cryptosystem involved.
>
> You may read that it would take "every computer in the world a million years to defeat encryption system *X*." Often, these claims are based on the sheer number of possible keys for a particular system.
>
> Advances in mathematics and computing power are making it possible to defeat more cryptosystems every year. Even systems that tout high-bit strength (128-bit, 1024-bit, and so on) have weaknesses unrelated to the size of the key.
>
> Ultimately, the security of any system is determined by its overall design and not by the strength of a particular key alone.

One application that uses PKC to protect your e-mail is Pretty Good Privacy (PGP). At the end of this chapter, I briefly discuss PGP and where you can download it.

Examining hybrid systems

Many modern cryptosystems use a combination of secret and public key cryptography. Because they only use one key, secret key systems tend to be faster than public systems. Because of this, a combination of both secret and public cryptography is often used in different networking cryptosystems.

The problem of key distribution is solved by using a public key system to encode and distribute the secret key, which is then used to secure further communications (see Figure 10-6). This technique is sometimes referred to as a *digital envelope.* This is common in financial networks, including Automated Teller Machine (ATM) systems, and in secure E-commerce systems.

Example applications of a hybrid system are digital signatures and digital certificates.

Figure 10-6: Key distribution with public key cryptography

Digital signatures

Digital signatures provide proof that a message hasn't been altered in transit and that it originated with the sender of the message. Digital signatures are produced with an encrypted one-way-hash of a message or binary file, combined with some application of public key cryptography (see Figure 10-7).

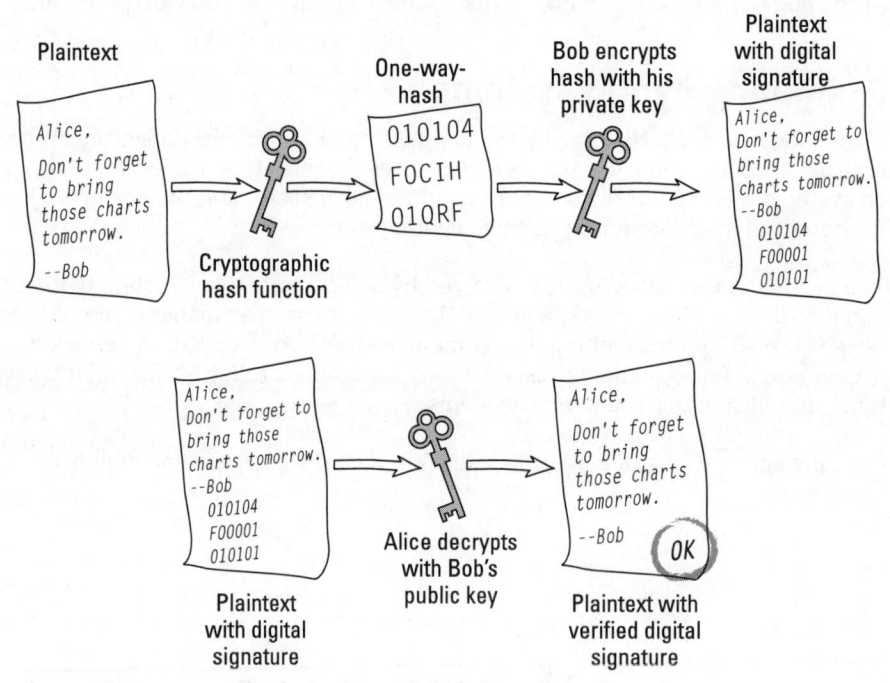

Figure 10-7: One method of using a digital signature

A one-way-hash function is an encryption technique that produces a sort of digital fingerprint, unique to the document (see Figure 10-8). It is one-way because you cannot derive the original message from the hash; you can only verify that the message is unaltered. Any change in the message changes the one-way-hash.

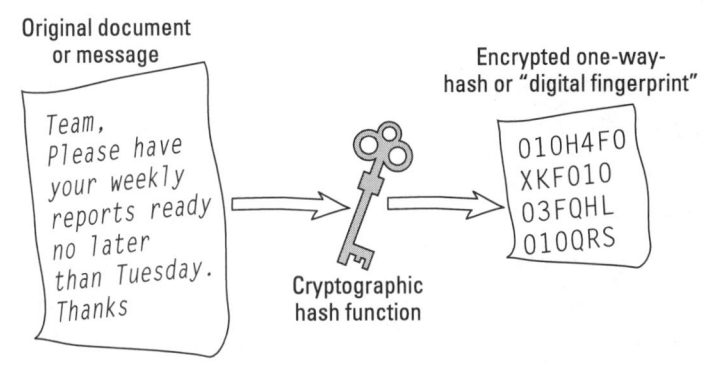

Figure 10-8: A one-way-hash of a message

While a digital signature proves that a message originated with a particular sender and that no one has altered it, it does not prove that the sender is who he says he is. If Alice receives a message with Bob's digital signature, she only knows that it originated from someone claiming to be Bob, and that it remains unaltered. She has no way of knowing that Bob actually sent it.

Verifying the identity of a sender requires an application of cryptography called a digital certificate.

Digital certificates

A digital certificate provides third-party verification that a party to a transaction is indeed who the party claims to be. In short, a digital certificate is a sort of digital I.D. card issued by a trusted third party called a *certificate authority*. One example of a certificate authority is VeriSign Corporation (www.verisign.com). VeriSign verifies the identity of an individual or organization and issues a digital certificate.

At one time or another, you have probably seen a warning about a digital certificate while surfing the Web (see Figure 10-9). These warnings indicate that your browser could not verify the identity of the Web site you are visiting. This is important, as crackers have set up fake Web sites that look like sites belonging to authentic companies, including Microsoft and Amazon.com. Usually these sites are identical to the legitimate site, but have a misspelled address (such as www.amzon.com). However, the crackers cannot forge the site's digital certificate. If you see a warning that a certificate is not authentic, especially if it is a major site, avoid providing personal information or credit card data.

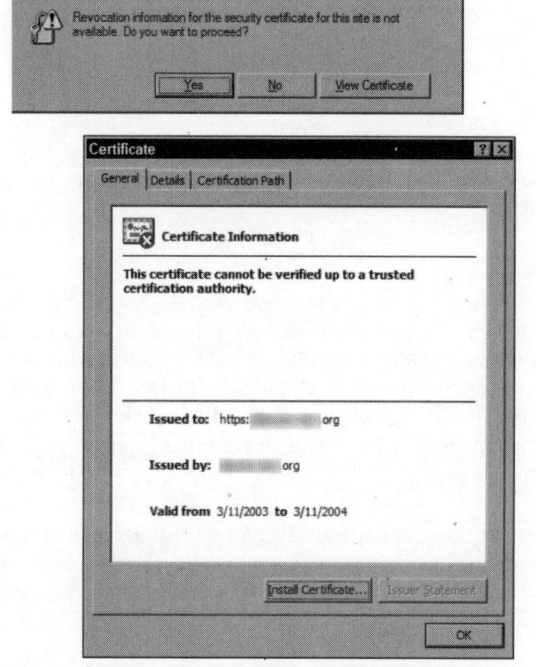

Figure 10-9: A digital certificate warning

Entropy and key strength

Advertisers like to stress key length when promoting the security of a particular cryptosystem. Usually this is expressed in bits, such as 128-bit or 64-bit encryption. Key length isn't an indicator of security; overall design of a cryptosystem and the entropy in the key are more important than the length of the key.

In a cryptosystem, *entropy* is the amount of disorder or randomness present in a key or in the phrase used to generate a key. Entropy represents the number of possible patterns contained within random data. If a key has low entropy, there are fewer discernable patterns within it, and that increases the likelihood that a pattern will be identified and the key discovered (this is oversimplified, but sufficient for your purposes).

Entropy is an important concept, one worth remembering. If you place all of your faith in an "unbeatable" 128-bit encryption system, then use an eight-character password to generate a 128-bit key, you've just undermined your security. Systems that use short passwords or pass phrases to generate keys typically do not create enough entropy in the key to make it very secure.

Cryptosystems use different methods to generate entropy in keys. Because it's impossible for a computer to create truly random numbers, researchers have devised ingenious ways of accomplishing this. Some methods sample environmental data (noise, temperature), while others collect random input from users (time between keystrokes, or mouse clicks).

When generating keys for your cryptosystem, always use long, nonsensical phrases or passwords to ensure that the keys you generate will contain sufficient entropy to remain secure.

Debunking Wired Equivalent Privacy (WEP)

Wired Equivalent Privacy (WEP) is the original encryption system created for 802.11 networks. To say WEP is flawed is like saying that Alaska can be a bit chilly; it doesn't do the enormity of the problem any justice. WEP is woefully insecure, and any determined person can defeat it using free tools available on the Internet.

Understanding why WEP fails

The WEP encryption standard has two primary weaknesses that make it susceptible to attack: key distribution and encryption. Both are seriously flawed and provide opportunities for crackers to defeat WEP encryption.

Crackers can passively attack WEP, meaning that they don't have to do anything overt to discover the key and perhaps reveal their presence. Because of the way

WEP implements encryption, a cracker can passively record network traffic, and once he has recorded a maximum of 25GB of data, recover the encryption key in seconds. On a busy WLAN, it's possible to record this amount of data in a few hours.

> **Note** Describing the operation of any encryption system in any depth is complicated and beyond the scope of this book. WEP, although flawed, is no exception so this is an extremely simplified explanation of how WEP actually works—for the sole purpose of illustrating some of the problems.

The big problem with key distribution in WEP is that the standard doesn't even describe how manufacturers should implement this feature. In practice, and on most consumer WLANs, everyone on the network is using the same key. This means that collecting traffic between any two nodes on the network will eventually reveal the key for everyone's communication.

WEP uses a random 24-bit string called an *initialization vector* (IV) combined with the secret key to create a *pseudo-random key* for each data packet. This is supposedly to prevent two ciphertexts from being encrypted with the same key. Remember, one of the ways to attack encryption is statistical analyses of patterns, such as frequency analyses.

The big problem with this approach is the way the WEP standard utilizes the IV. The IV is sent with each packet, as plaintext, that is in-the-clear. Twenty-four bits is short; if a cracker collects enough data packets he will eventually find two packets, or ciphertexts, generated with the same initialization vector. Then, using statistical analysis, he can discover the secret key and decipher all communications.

The WEP standard doesn't establish how manufacturers should generate the IV or that it must be 24 bits long. Do you remember what I said about entropy, or randomness of an encryption? The shorter the IV, the greater likelihood that the same number will be used more than once. Worse than this, some manufacturers do not even change the IV packets, because according to the WEP standard, it's optional.

Because WEP implements encryption so poorly and doesn't require manufacturers to comply with portions of the standard (key length and generation), it's susceptible to basic cryptanalysis techniques and unsuitable for protecting important data.

Why you should still use WEP

Now that you understand why WEP is insecure, I must stress one point: if WEP is all that's available on your Wi-Fi equipment, use it. Even flawed encryption is better than none at all. As long as you realize that WEP affords you minimal protection and protect your data accordingly, then there is no reason not to use it.

Even though WEP is easily cracked, you'll be adding a step for anyone who wants to break into your WLAN, and that may make a cracker move on to try another network rather than make the effort.

If you use WEP on your network, there are a few things that you can do to avoid making it any easier for crackers. These are:

- Upgrade your device's firmware
- Use passwords or phrases that are as long as possible
- Select the highest encryption setting
- Change the factory default key
- Change the key periodically

Most new devices have upgradeable firmware. If your device does, be sure that you download and install any updates provided by the manufacturer. In many cases, the manufacturer may have improved its implementation of WEP, making it slightly more secure. In some cases, manufacturers have replaced WEP with the improved Wireless Protected Access (WPA) standard.

If your software prompts you for a password or passphrase for key generation, use one that is as long as possible. Follow all the best practices for passwords listed in Chapter 11. When using a keyphrase, don't select a passage from a book. This is the equivalent of using a password comprised of a single dictionary word.

Many Wi-Fi devices offer encryption settings as low as 40 bits and as high as 128 bits. Use the highest setting available on your device. This will increase the difficulty of discovering the WEP key, although the IV remains 24 bits and is still a weak point.

If your device has a default WEP key, be sure to change that prior to setting up your WLAN. Any default settings on your Wi-Fi gear are common knowledge to crackers and as a result are completely insecure.

Lastly, change your WEP key periodically. The more often you change it, the harder it will be for a cracker to discover the key and compromise your network.

Investigating Wi-Fi Protected Access (WPA)

In reaction to the problems that researchers discovered about the WEP encryption protocol, the Wi-Fi Alliance began work on the next generation wireless encryption standard. Called Wireless Protected Access (WPA), this standard corrects many of the problems associated with WEP and actually provides a degree of protection to wireless networks.

The Wi-Fi Alliance based WPA on an early, unpublished subset of the IEEE 802.11i security standard. Since then, the IEEE has published the final version of the 802.11i standard, and subsequent versions of WPA will incorporate more of 802.11i's features.

Most new Wi-Fi gear supports WPA rather than WEP encryption. Some older devices are upgradeable to WPA simply by downloading a firmware update from the manufacturer. If WPA is available as an option on your Wi-Fi device, use it. It's far more secure than WEP and will go a long way toward securing your home WLAN.

That said, WPA isn't invulnerable to attack. To protect access points from attack, WPA authenticates users. If WPA detects two packets of unauthorized data within a one-second period, it assumes that it's under attack and shuts down. Although this feature is meant to thwart an attack, it's easily exploitable as a means to a denial-of-service (DoS) attack.

All an attacker has to do is send two bad packets every minute, and the access point will continue to shut down and reboot (see Figure 10-10). In fact, there are already attack tools available on the Web that automate this and make it easy to launch a WPA DoS attack.

In addition to DoS attack vulnerability, crackers can also attack WPA encryption. However, unlike WEP, the weaknesses in WPA encryption aren't inherent in the WPA standard, but in the manner that some manufacturers have implemented the interface.

Some devices have an interface that limits the length of the password or passphrase a user can use to 20 characters or less. If you type words found in a dictionary or use a short passphrase, a cracker can use any one of a number of password-cracking programs to recover the password and discover the WPA key.

Once again, this isn't a problem with WPA, but an interface problem. For maximum security, use long passwords or phrases with random letters and numbers. Ideally, you should enter hexadecimal numbers if your software allows you to.

Chapter 10 ✦ **Encryption and Wi-Fi** 183

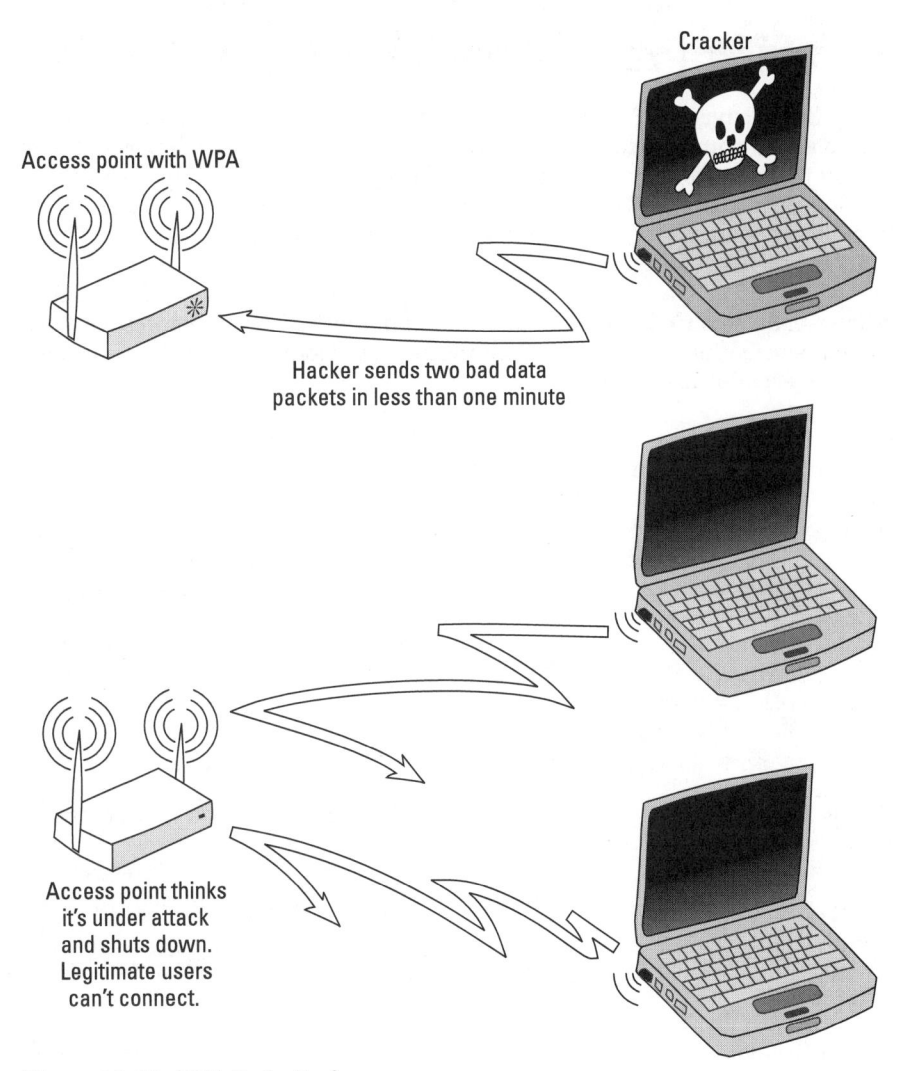

Figure 10-10: WPA DoS attack

Introducing Pretty Good Privacy (PGP)

Pretty Good Privacy (PGP) is an encryption program used by millions of people to secure e-mail communication and files. PGP uses a secret key to encrypt the message of the file and then uses public key encryption to encrypt the secret key, which is sent with the file to its destination (see Figure 10-11). PGP users can select from different encryption algorithms when encrypting their messages. PGP also supports the use of digital signatures.

PGP is available on Windows and Macintosh platforms and integrates with mainstream e-mail applications, including Microsoft Outlook, Outlook Express, Eudora, Entourage, and Apple Mail.

> **On The Web**
> You can purchase and download PGP Personal Desktop at www.pgp.com, or a freeware version from www.pgpi.org.

Figure 10-11: Using PGP to encrypt a message

Summary

In this chapter I introduced encryption, explained some of the concepts involved, and introduced to types of encryption used on WLANs. The topics I covered include:

- Explanation of encryption terms
- Modern encryption techniques and applications
- Wired Equivalent Privacy (WEP) encryption
- Wi-Fi Protected Access (WPA) encryption
- Introduction to Pretty Good Privacy (PGP)

Encryption can play an important part in securing your data. When used properly, WEP and WPA will help secure your WLAN and keep your communications secure.

✦ ✦ ✦

Securing Your WLAN

CHAPTER 11

In This Chapter

Understanding your WLAN vulnerabilities

Checking dangerous default settings on your network

Setting up a personal firewall

More things you can do to protect yourself

You've no doubt reviewed all of the threats that your Wi-Fi network may face, now I will discuss the steps you can take to secure it. It is possible to lock down a Wi-Fi network, at least to the point that it would be very difficult for anyone but a technically proficient cracker to gain access. It's also possible to make it harder for crackers and wardrivers to locate your network in the first place. This chapter lays out all of the steps you can take to protect your network.

Security is a balancing act between risk and the amount of effort (and inconvenience) you're willing to expend to protect yourself. If your stance regarding security is too radical, you may alienate network users or customers who may try to circumvent your security measures just to get their jobs done with less hassle. After a while, you may even find yourself taking shortcuts if you've set the bar too high or created a tedious security program.

Take the time to weigh the risks against the impact that security will have on network users and your ability to get your work done. The steps presented in this chapter should provide an adequate level of security for most small home or office networks without adversely effecting users and ultimately undermining your security program.

Understanding WLAN Vulnerabilities

When securing your wireless network, begin by assessing your current situation and weighing the possible risks to your network against the cost or inconvenience of the steps that you can take to lock it down. Although every WLAN is vulnerable to some degree, some are more likely to be targeted than others due to location or the way the network is designed.

If your wireless network is located in an urban area, particularly in a location where there is a large tech-savvy population (Silicon Valley, for example), it's more likely to be discovered by wardrivers or crackers than it would be in a rural area.

Maybe you think that you don't need to secure your wireless network because you feel that you have nothing sensitive or valuable to secure. Even if this were the case — which it most likely isn't — you should realize that your wireless network may present another risk. If your WLAN connects to an existing wired Ethernet LAN, it puts the entire wired network segment at risk.

Radio waves are broadcast in all directions and are easily intercepted. It's difficult, if not impossible, to control where your Wi-Fi signal goes and where it doesn't go. The open nature of the network media makes wireless LANs less secure than wired networks.

Changing Dangerous Default Settings

The majority of Wi-Fi hardware is user friendly. Most people can set up an access point and Wi-Fi cards even if they are networking novices. This user friendliness has helped to drive adoption of Wi-Fi networks; but it may also leave many WLANs vulnerable unless users take steps to secure them.

Users create some of the biggest security problems when they set up their wireless networks and fail to change the default settings. These defaults, including SSID, IP subnets, and administrative usernames and passwords, are well known to wardrivers and crackers, and are available on the Internet. Failing to change these settings invites disaster.

On The Web For lists of known default settings for Wi-Fi equipment visit www.cirt.net.

Taking the time to follow the steps in this chapter will make your WLAN reasonably secure, and will deter casual wardrivers and inexperienced crackers. However, a skilled and determined cracker can still get around some of these measures, so you'll need to be diligent in checking your security periodically to make sure.

Rethinking the SSID

Every Wi-Fi access point has an assigned Service Set Identifier (SSID) that identifies the network. Leaving the default SSID in place when you set up your WLAN can help crackers or wardrivers identify which hardware you are using. A wardriver or cracker can discover the SSID of your network using an application like NetStumbler (see Figure 11-1). For example, the default SSID for most Linksys hardware is "Linksys," for Netgear access points it's "netgear." These default SSIDs clearly identify the hardware. Once a cracker knows that, he can look up the default administrative usernames and passwords, allowing him to compromise the network with minimum effort.

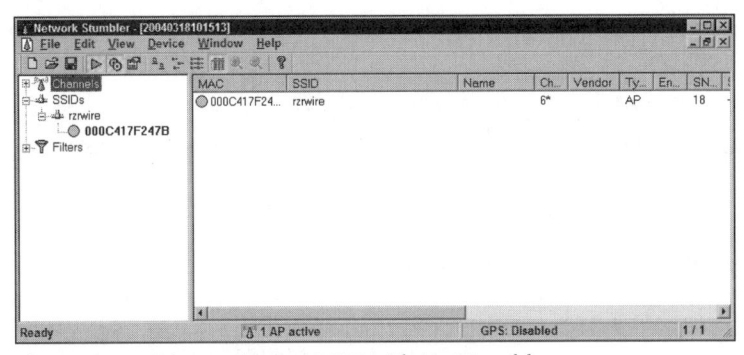

Figure 11-1: Discovering an SSID with NetStumbler

Fortunately, you can change the SSID on your access points at any time (see Figure 11-2). You just have to be sure to change the SSID on your wireless adapters as well. When you are change the default SSID, avoid using something that identifies your residence or business. Some of the best practices for changing your SSID are:

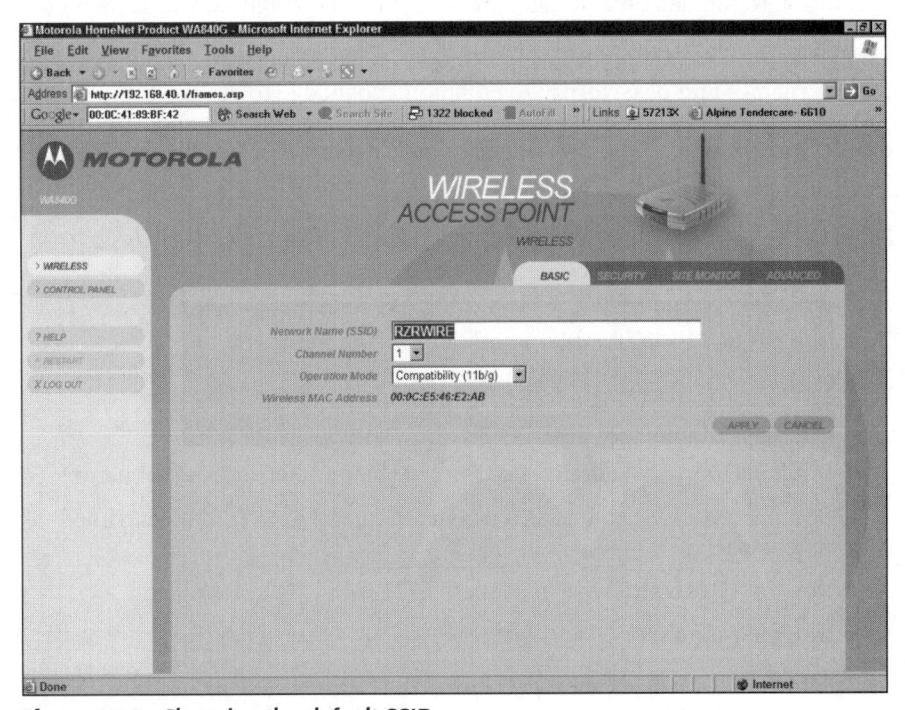

Figure 11-2: Changing the default SSID

- Never use personal information in your SSID (name, birth date, phone number, or street address).
- Don't use any of your usernames or passwords.
- Don't use an SSID that identifies your hardware (for example, Linksys BEFW11S4).
- Create long SSIDs that are composed of both letters and numbers.
- Don't use an SSID that identifies the access point's location or dept (for example, "accounting").

Note: The SSID must be all caps for MS-DOS clients to be able to recognize it.

Changing the default SSID is the first and easiest step to deterring casual wardrivers and inexperienced crackers from trespassing on your WLAN.

Changing passwords and usernames

In addition to the default SSID, all hardware comes configured with a default administrative username and password. It's especially important that you take the time to change these settings (see Figure 11-3) because, like the SSID, default administrative usernames and passwords are common knowledge to crackers and wardrivers and are available on a number of Web sites. If you neglect to change the settings, a cracker can determine which hardware you are using and use the default username and password to take over the access point and compromise your WLAN.

Some of the best practices for choosing a username and password are:

- Use passwords that are at least six characters long.
- Use passwords that contain both letters and numbers.
- Use a mix of upper- and lowercase letters.
- Don't use common names, places, or "dictionary" words.
- Don't use personal data as a password (birth date, address, or name)
- Don't use your username as a password in any form (doubled, tripled, or reversed).
- Don't use keyboard keys in sequence ("12345", "QWERTY", etc).
- Don't use acronyms.
- Don't reuse your old passwords.
- Don't write down passwords.
- Change your password frequently.

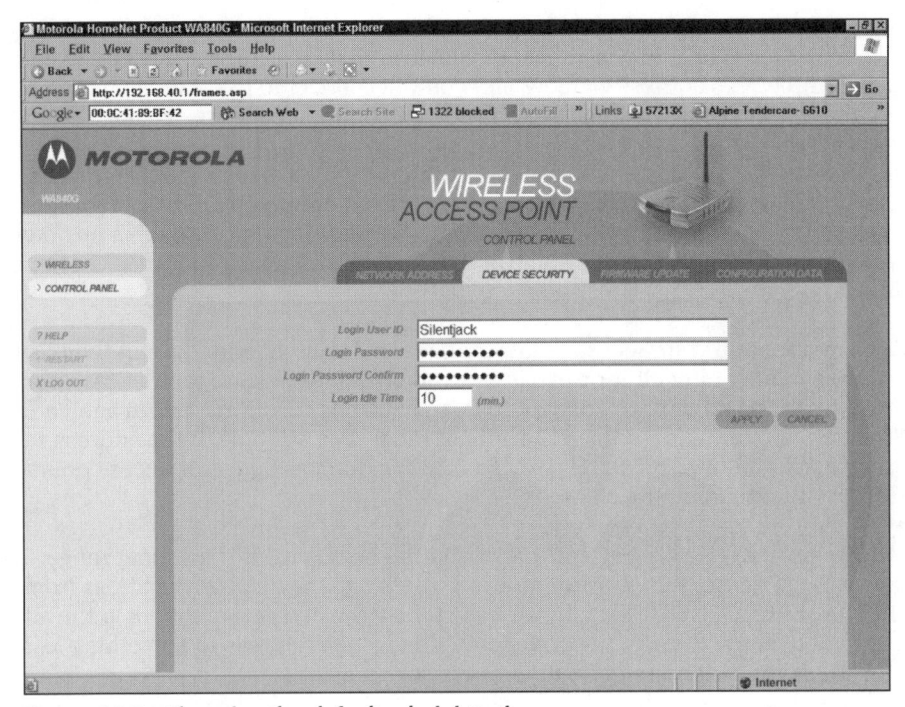

Figure 11-3: Changing the default administrative username

Using password-cracking utilities and any desktop computer, a cracker can discover passwords created with English or foreign dictionary words in a matter of seconds. During audits of some networks, I have been able to recover as many as 90 percent of users' passwords in less than one minute. Most had ignored commonsense best practices when choosing their passwords.

Editing IP addresses

Any device connected to the Internet or to a routed TCP/IP network (like a WAN) needs to have an IP address assigned to it in order for it to connect to the network. Wi-Fi access points have a default Internet Protocol (IP) address assigned by the manufacturer, as well as a default IP subnet and gateway address.

This can be a problem because, like everything else, the default network settings on access points are common knowledge. For example, the default network settings on a Motorola access point are:

✦ IP address: 192.168.40.1
✦ Default gateway 255.255.255.0

Almost all access points available have a Web interface that allows you to configure the device through a Web browser. All you need to do is enter the device's IP address in your browser's address window, and you will connect to the access points built-in Web server, which will display a login and configuration page (see Figure 11-4).

If a cracker knows the default network settings for a particular manufacturer's equipment, and he identifies which type of access point you are using (that is, through SSID broadcast), he can connect to the access point through a browser. If you haven't changed the username and password, he can log into the access point and take control of it.

When you set up your access point, change the default IP address and the IP subnet to make it more difficult for trespassers and crackers to guess your access point's IP address and attempt to connect to it (see Figure 11-5). Don't assign an IP address from the first few addresses in the subnet range. For example, if the address range of your WLAN is 192.168.0.0 to 192.168.255.255, don't assign your access point 192.168.0.1 or something close.

Access points typically have an address in the beginning of the subnet range, and any cracker knows this. For example, if you change the default IP address from 192.168.0.1 to 192.168.0.3, a cracker can still try the first few numbers in the address range and he will find your access point. Change the address to something less obvious to make it more difficult for a cracker to guess.

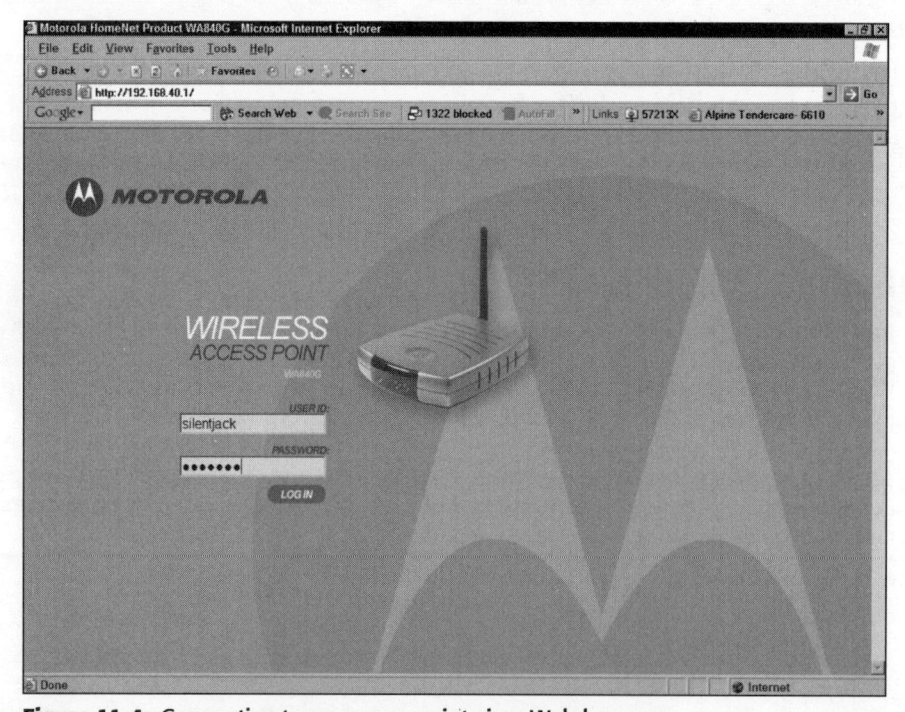

Figure 11-4: Connecting to an access point via a Web browser

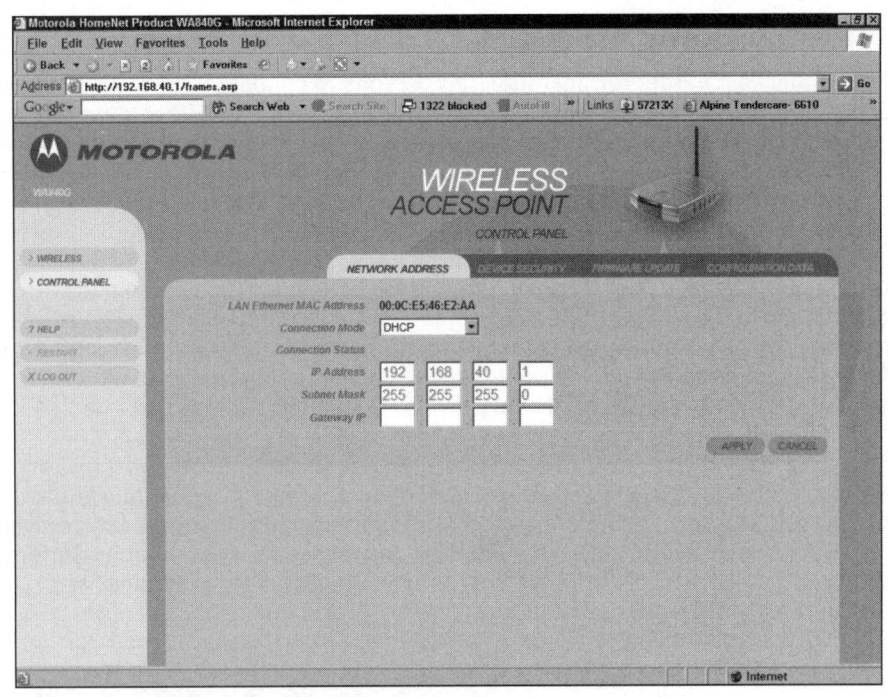

Figure 11-5: Changing network settings on an access point

Understanding DHCP

Most access points also act as routers, and many of these have Dynamic Host Configuration Protocol (DHCP) functionality built into them. Earlier in this chapter, I mentioned that every computer connected to a routed network or the Internet must have an IP address assigned to it.

DHCP simplifies this by automatically assigning IP addresses to network clients from the network's assigned address range. Each time a computer reboots or logs on to the network, the DHCP server (or the access point) assigns it an IP address. This is simpler than manually assigning an IP address for each client, especially on large networks. It also allows Internet Service Providers (ISPs) to recycle IP addresses among dial-up customers. Each time a customer dials in, the ISP assigns a new IP address. When the user disconnects, the IP address is recycled and assigned to another user.

Depending on your network configuration, you may want to disable the DHCP services on your access point. If you have a broadband modem with a router that has DHCP services, or your WLAN is part of a larger network with a DHCP server, these devices will assign IP addresses to your clients and access point when they connect.

Understanding why DHCP can be a problem

You can make it harder for unauthorized persons to connect to your network by disabling DHCP on your WLAN. An intruder who knows the SSID of your wireless network can configure his Wi-Fi card and attempt to connect to it. If you have DHCP enabled, your access point may assign the intruder an IP address and allow him to connect to the network and access network resources such as your Internet connection or printers.

Disabling DHCP services will prevent this from happening, and combined with the other steps in this chapter, will deter an inexperienced intruder (see Figure 11-6). However, if you do this, you will have to configure each network client with a static (unchanging) IP address. Each time you add a client to the network, whether it is a computer, printer, or other network device, a static IP address must be assigned to it.

If you are running Windows Internet Connection Sharing (ICS), you must disable ICS to turn off DHCP. ICS has an integrated DHCP server, and there is no way to disable the DHCP service without turning ICS off. If you have an access point or router providing DHCP service, refer to the help file or user manual for your device for instructions for disabling DHCP.

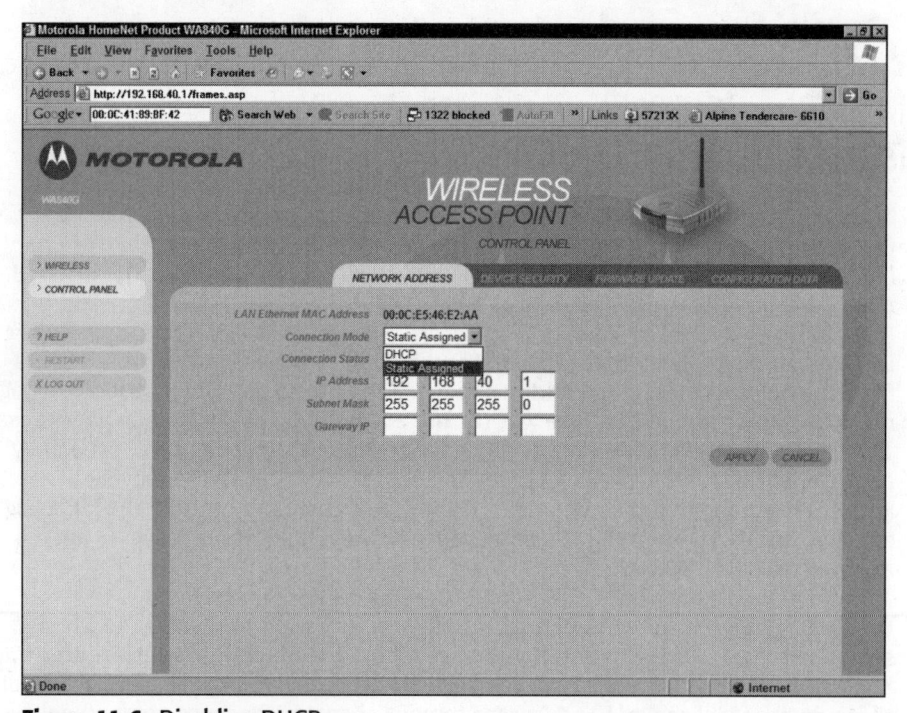

Figure 11-6: Disabling DHCP

Assigning static IP addresses

Assigning a static IP address to each of your wireless clients isn't difficult. Chances are good that whatever wireless adapter you are using came with configuration software that will enable you to do this with minimum difficulty (see Figure 11-7).

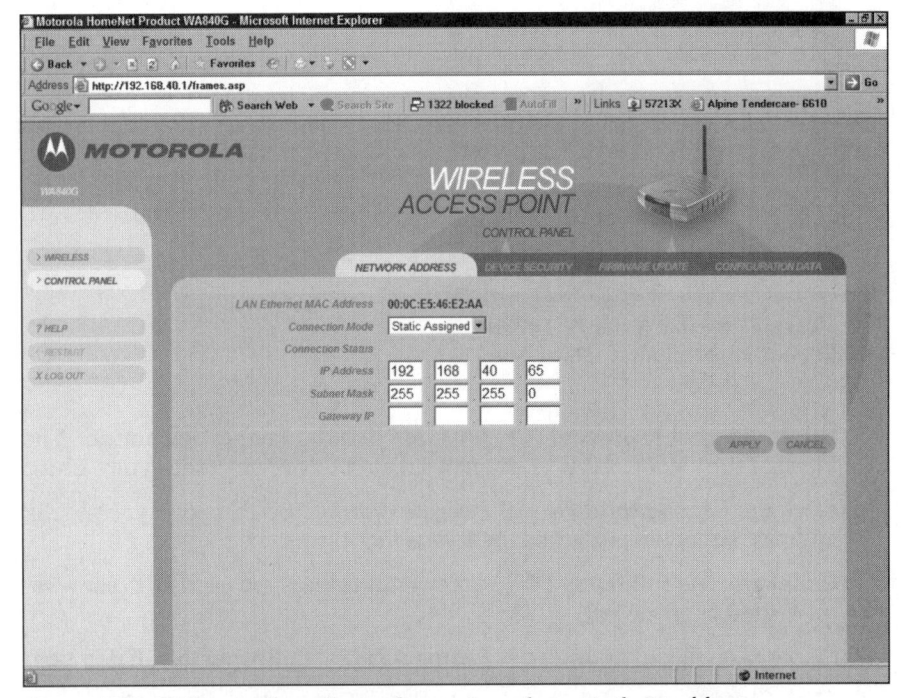

Figure 11-7: Using configuration software to assign a static IP address

If you have to configure your adapter manually, you can follow the steps below to assign a static IP address to a Windows XP client.

STEPS: Assigning a static IP address to a Windows XP client

1. **Left-click on the Start button.** The Windows XP Start menu appears.

2. **Click Control Panel on the Start menu.** The Network and Internet Connections window opens.

3. **In the Control Panel, double-click on the Network Connections icon.**

4. **Right-click on your wireless network connection, and click on Properties in the pop-up menu.** The Wireless Network Connection Properties dialog box opens (see Figure 11-8).

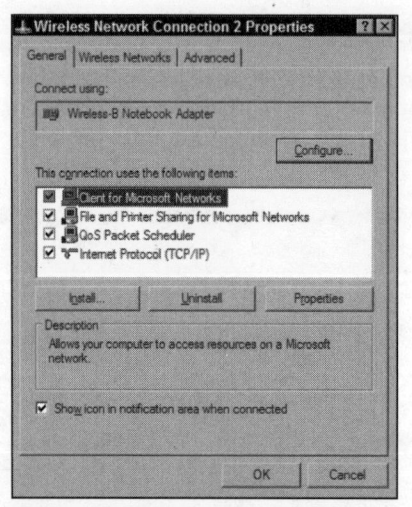

Figure 11-8: The Wireless Network Connection Properties dialog box

5. **Select Internet Protocol (TCP/IP), and click the Properties button.** The Internet Protocol (TCP/IP) Properties dialog box opens (see Figure 11-9).

6. **Click the Use the following IP address option.** You can now type your IP address, subnet mask, and default gateway.

7. **Under Use the following DNS server addresses, type your DNS servers (provided by your ISP).**

8. **Click OK to close the Internet Protocol (TCP/IP) Properties dialog box.**

9. **Click Close to close the Wireless Network Connection Properties dialog box.**

You will have to repeat these steps with each of your network clients, and you must use the configuration software that came with your access point or log into the access point's Web interface to assign it a static IP.

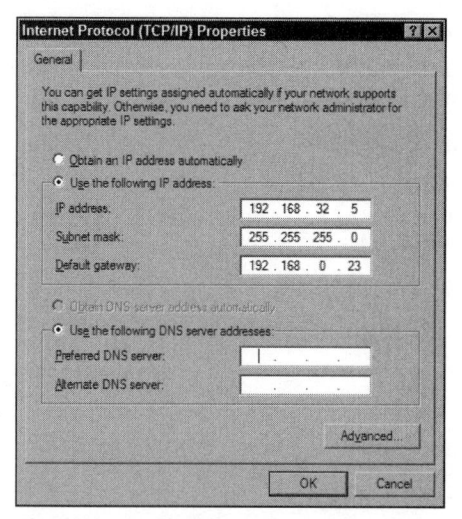

Figure 11-9: The Internet Protocol (TCP/IP) Properties dialog box

Filtering Network Traffic

If your access point has a firewall built in, another method of preventing intruders from gaining access to your WLAN is to filter your network's traffic. Filtering network traffic allows you to configure your firewall in a way that excludes all users except those that your configuration allows to connect.

You can use one of two methods to filter your network traffic. One method uses the Media Access Control (MAC) address; the other filters by IP address (see Figure 11-10). Either way, filtering only allows known (and approved) addresses to associate with your access point. Manufacturers encode each network adapter and device with a unique alphanumeric MAC address. In most cases, the MAC address is permanent, but crackers can modify the MAC on some devices.

Authorized MAC addresses

MAC 01-0F-22-6E-13-8D
MAC 00-0C-41-7F-24-7C

Authorized IP addresses

IP 192.168.0.12
Thru
IP 192.168.0.44

MAC 00-0C-41-7F-24-7C
IP 192.168.0.12

Access point allows authorized IP and MAC addresses to connect and denies addresses that aren't on lists.

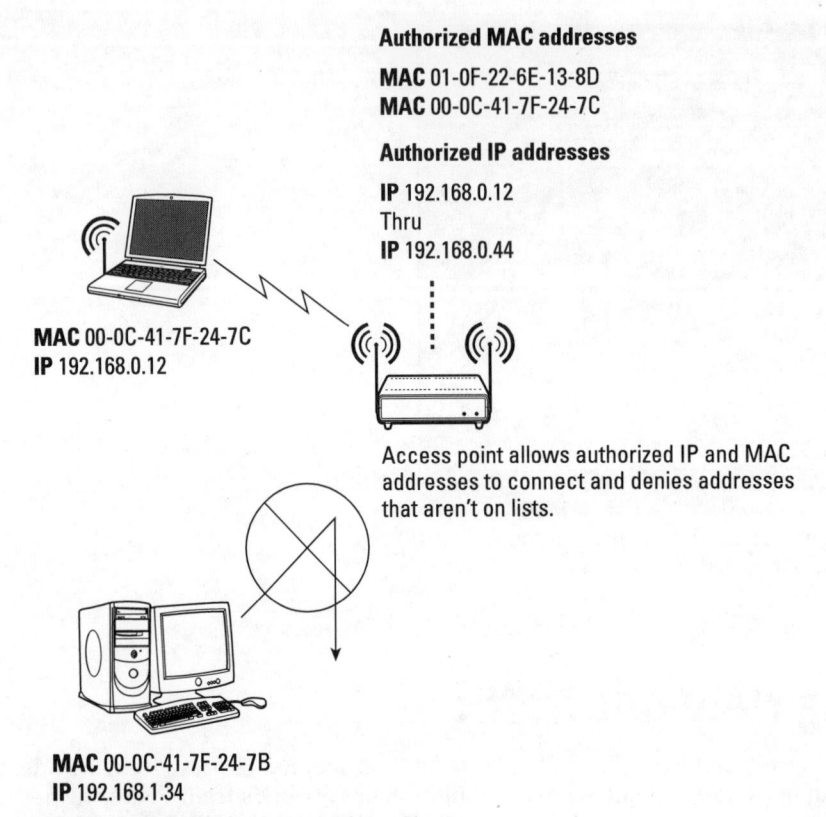

MAC 00-0C-41-7F-24-7B
IP 192.168.1.34

Figure 11-10: MAC and IP address filtering

Because of this, filtering isn't foolproof. A cracker can use a wireless sniffer and capture network data packets to analyze. Network data packets contain routing information that includes both the IP address and MAC address of the sending station. Using this information, a cracker can configure his adapter to use the MAC (or IP) address of a legitimate user and thwart your firewall's filter (see Figure 11-11).

Although it's not foolproof, analyzing data packets to learn MAC addresses or legitimate IP addresses raises the bar high enough so that only technically proficient (and determined) intruders are able to access your WLAN. Because the majority of would-be intruders aren't that skilled and unlikely to waste much time trying to beat it, filtering is an effective additional step to securing your wireless network.

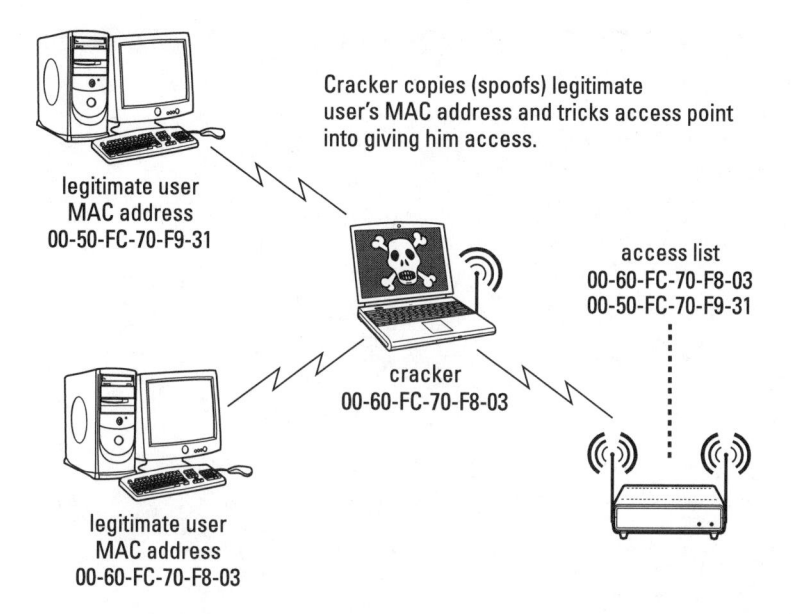

Figure 11-11: Spoofing a MAC address to get past MAC filtering

Activating MAC address filtering

The exact steps to implement MAC or IP filtering depend on the hardware you are using and are different for every brand of access point or firewall. The most likely scenario is that you can set up filtering through the configuration software or Web interface for your device (see Figure 11-12).

To set up a list of MAC addresses that are allowed to associate with your access point, gather the MAC address of every wireless adapter on your network. The easiest way to learn the MAC address of each of your adapters is to use the configuration software that came with the adapter. However, you can also find the MAC address of your adapter in Windows XP or Windows 2000 by following these steps:

Part II ✦ Protecting Yourself

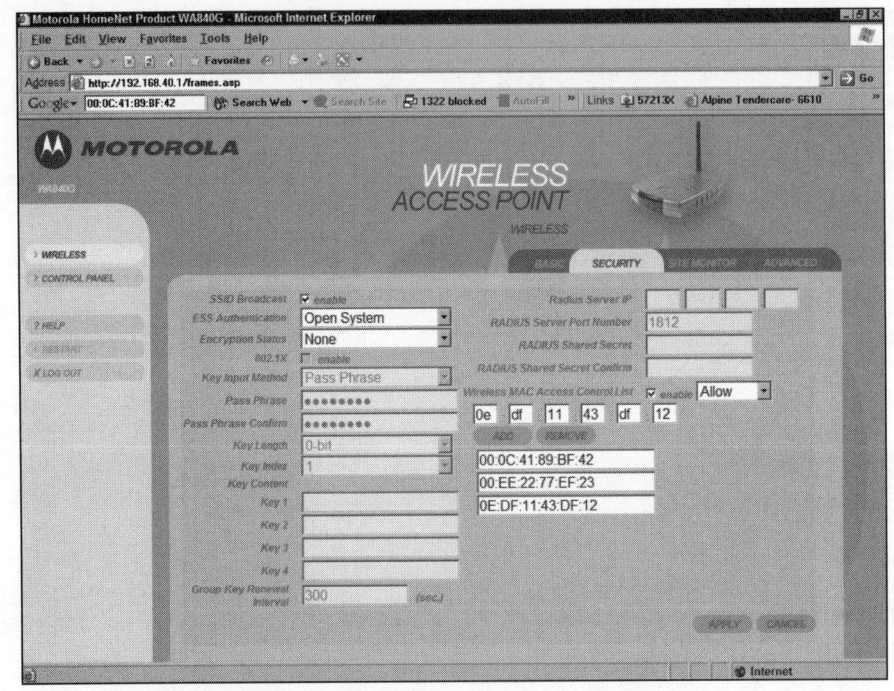

Figure 11-12: Implementing MAC filtering through a Web interface

STEPS: Finding the MAC address of your adapter in Windows XP/2000

1. **Left-click the Start Button.** The Start menu appears.

2. **In the Start Menu, click Run.** A small Run dialog box opens (see Figure 11-13).

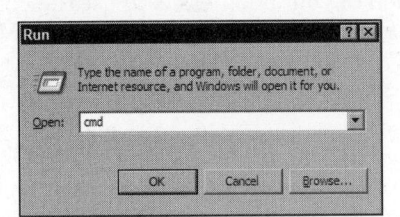

Figure 11-13: The Run dialog box

3. **In the Run dialog box, type** CMD **and click OK.** A command window appears.

4. **In the command window, type** ipconfig /all. There must be a space after `ipconfig`. The command window displays the network configuration for your adapter, including IP and MAC address (see Figure 11-14).

```
C:\>ipconfig /all

Windows IP Configuration

        Host Name . . . . . . . . . . . . : Mr-Krabbs
        Primary Dns Suffix  . . . . . . . :
        Node Type . . . . . . . . . . . . : Mixed
        IP Routing Enabled. . . . . . . . : No
        WINS Proxy Enabled. . . . . . . . : No

Ethernet adapter Wireless Network Connection 2:

        Connection-specific DNS Suffix  . : mshome.net
        Description . . . . . . . . . . . : Wireless-B Notebook Adapter
        Physical Address. . . . . . . . . : 00-0C-41-89-BF-42
        Dhcp Enabled. . . . . . . . . . . : Yes
        Autoconfiguration Enabled . . . . : Yes
        IP Address. . . . . . . . . . . . : 192.168.0.166
        Subnet Mask . . . . . . . . . . . : 255.255.255.0
        Default Gateway . . . . . . . . . : 192.168.0.1
        DHCP Server . . . . . . . . . . . : 192.168.0.1
        DNS Servers . . . . . . . . . . . : 192.168.0.1
        Lease Obtained. . . . . . . . . . : Thursday, March 18, 2004 10:29:17 AM
        Lease Expires . . . . . . . . . . : Thursday, March 25, 2004 10:29:17 AM

C:\>
```

Figure 11-14: The command window displaying IP and MAC addresses

Repeat these steps to collect the MAC address for each of the adapters that you allow to associate with your access point.

Implementing IP address filtering

Like MAC address filtering, the steps you take to configure your access point or firewall to allow certain IP addresses (or a range of them) to associate with your access point is dependent on the device that you are using. Again, the most likely way for you to accomplish this is with the configuration software that came with your device (see Figure 11-15).

To collect the IP addresses of wireless adapters that you allow to associate with your access point, follow the same steps you used to gather the MAC addresses. The `ipconfig /all` command will also display the adapter's IP address (see Figure 11-16).

Caution If you are using DHCP to supply your network clients with dynamic (changing) IP addresses, don't filter based on IP addresses. Any IP addresses you collect now will change when you reboot your computer or disconnect and then reestablish your network connection.

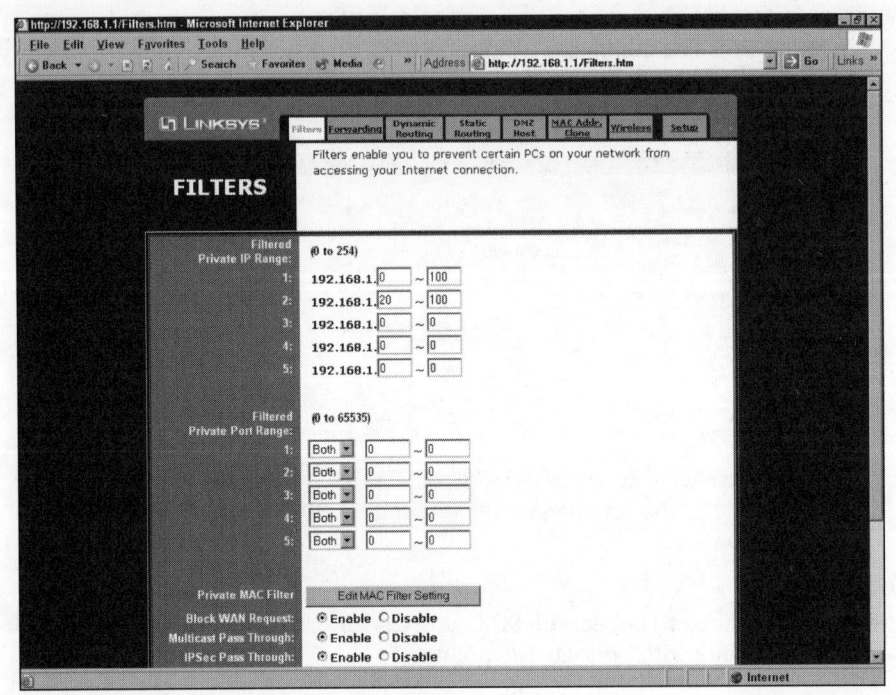

Figure 11-15: Configuring IP address filtering

Figure 11-16: The command window displaying a wireless adapter's IP address

Modifying Broadcast Parameters

Modifying the broadcast parameters of your access point to limit the distance your network's radio signals travel is another way to increase the security of your WLAN. Limiting the signals makes it harder to detect network settings.

Adjusting the power

For security, you can decrease the power of the radio signal that your access point broadcasts to limit propagation of radio waves outside of your home or office. This can reduce the chances that crackers or wardrivers will detect the signal.

Unfortunately, most common access points manufactured for home or small-office use do not have a built-in way to reduce the broadcast power. If this is the case with your access point, and you want to reduce its power output, you can use inline attenuators.

Inline attenuators are devices that attach to your access point and reduce the signal output by creating resistance and absorbing some of the output power. Most RF attenuators come with standard connectors, and you install them between the access point and the antenna (see Figure 11-17).

You need to do a site survey to be sure that the signal from your access point still reaches the areas of your home or office that you need it to. The easiest way to do this is to use a portable wireless client like a Wi-Fi-equipped PalmOS or Pocket PC handheld PDA, or a Wi-Fi-equipped laptop computer.

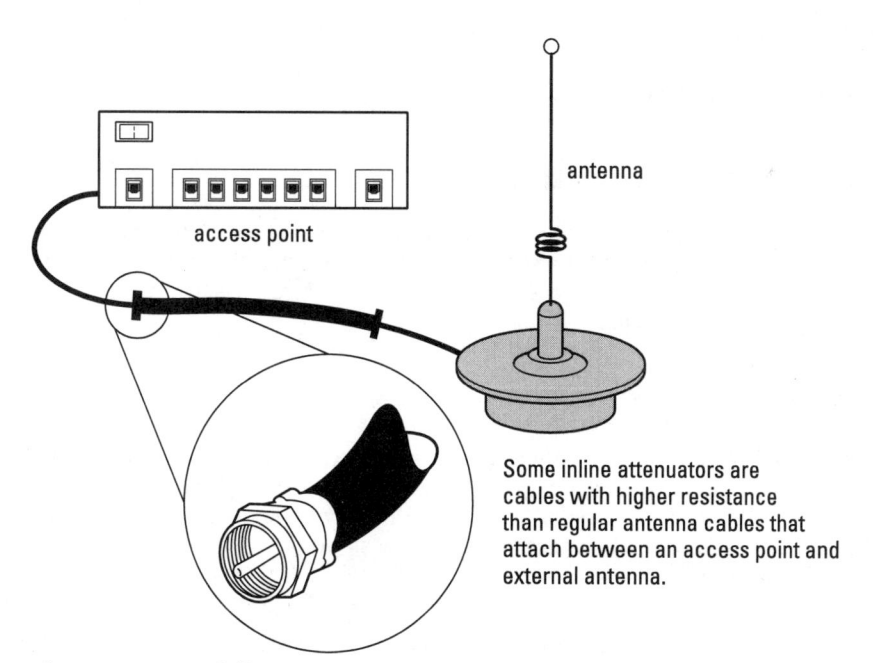

Some inline attenuators are cables with higher resistance than regular antenna cables that attach between an access point and external antenna.

Figure 11-17: An inline attenuator

For software, you can use any of the free wireless stumblers or sniffers available online. You may also be able to use the configuration software that came with your Wi-Fi adapter if it has a site-survey mode (see Figure 11-18). Walk around your home or office with the wireless client, and make a note of where you can and cannot receive the access point's signal. Make sure that by reducing the power output of your access point that you don't prevent your own network clients from accessing the WLAN.

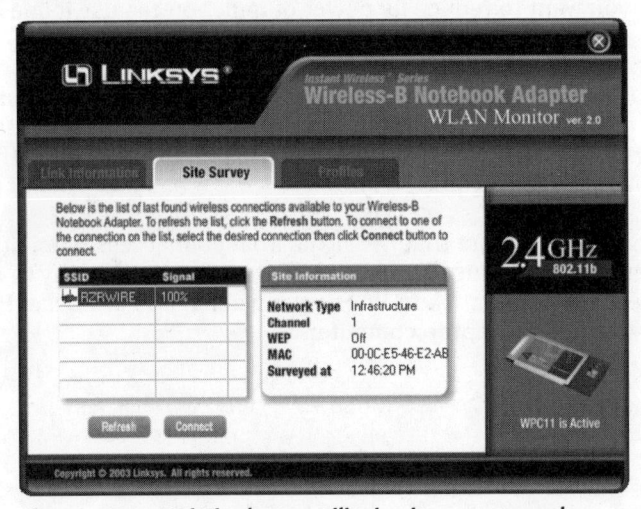

Figure 11-18: Wi-Fi adapter utility in site-survey mode

Note There are also small handheld Wi-Fi access point detectors available. I've had mixed results with these devices. They aren't a reliable way to determine if intruders can detect your WLAN's signal because they aren't as sensitive or reliable as a wireless laptop or PDA.

Turning off SSID broadcast

By default, every access point announces itself to the world by broadcasting its SSID. Every few seconds the access point broadcasts a data packet known as a beacon frame. The beacon frame contains the SSID. Any Wi-Fi client can detect the SSID broadcast and attempt to connect with the access point.

In addition to changing the default SSID name, you can also disable the SSID broadcast on most access points. To do this you have to use the access point's configuration utility or the Web interface.

When combined with the other security steps in this chapter, turning off the SSID broadcast helps conceal your WLAN from casual wardrivers or crackers. However, even if you disable the SSID broadcast, a cracker can still discover the SSID. Beacon frames aren't the only data packets that contain the SSID, and a determined cracker can capture and analyze network traffic to discover the SSID. However, it takes extra effort to do so, and disabling the SSID broadcast will hide your access point from many wardrivers.

Setting a minimum connection speed

On many access points, you can set a minimum connection speed for any client attempting to associate with the access point. Because the connection speed drops the farther you are from the access point, the effect is similar to reducing the broadcast strength of the access point.

Presumably, wardrivers and other persons located outside of your home do not have a signal that is strong enough to enable them to connect at the minimum speed you decide on. Conduct a site survey to determine if the signal is strong enough for your own computers to connect at the minimum speed.

Using Encryption

You can encrypt data to prevent wireless trespassers from eavesdropping and stealing your information. The two main types of encryption available on Wi-Fi gear are Wired Equivalent Privacy (WEP) and the newer Wi-Fi Protected Access (WPA). Encrypting your data can prevent crackers from stealing passwords or sniffing MAC addresses.

All Wi-Fi gear has encryption capabilities built in. Consider using it if you transfer a lot of sensitive data between computers or are in a densely populated urban area where you are more likely to have crackers or wardrivers attempting to intercept your signal.

Using encryption adds a little overhead to your network. Encrypting and decrypting network traffic takes more processing power and may slow down some older Wi-Fi gear. However, the effect should be minimal, and the security benefits outweigh the effect on WLAN performance.

Using WEP

Wired Equivalent Privacy (WEP) is the original encryption method used by Wi-Fi networks. It's available on all Wi-Fi hardware, and you should consider enabling it to add an extra layer of security to your network traffic. Once enabled, WEP requires very little input from you, other than selecting passphrases for key generation.

> **Note** Because skilled crackers can defeat WEP easily, it is ineffective for defending against them. Still, WEP is a sufficient deterrent to most would-be crackers and you should consider using it anyway (unless WPA is available as an option). WEP can serve as an intellectual firewall, making it difficult for casual wardrivers or unskilled crackers to access your WLAN. Most will simply move on to the next, less secure target.

You can activate WEP with the configuration software that came with your access point and adapters. When given a choice, always choose open authentication instead of shared key authentication. Crackers can defeat shared key authentication more easily than they can defeat open authentication.

Installing WPA firmware upgrades

WPA is the newer, more secure option for encrypting your WLAN. WPA is available on most new access points and adapters, particularly 802.11g and 802.11a devices. WPA is available for some older devices, such as 802.11b, as a firmware upgrade. Not all devices are upgradeable; you'll have to check with the manufacturer of your device. The easiest place to do this is at the manufacturer's Web site.

Note WPA isn't backward compatible with WEP, so if you're going to use it, all of the devices on your network must have WPA capabilities.

Firmware upgrades are executable files that you run to update the software in a hardware device's flash memory. This is also known as flashing a device. For added safety, download executable firmware upgrades directly from the manufacturer's Web site, not from third-party sites, bulletin boards, or newsgroups. It is easy for a cracker to disguise malicious software as a firmware upgrade, so always use caution when obtaining executable files.

If your devices have WPA capability, you can enable them with the configuration software supplied with your adapter or access point. Although it is more secure than WEP, WPA isn't impervious and is vulnerable to some attacks. Still, WPA is an improvement over WEP, and I recommend that you install and activate WPA encryption on your Wi-Fi network if possible.

A WPA Weakness

Recently, researchers discovered a weakness in WPA on consumer Wi-Fi networks that allows a cracker to recover the encryption key and defeat WPA. This is not due to a flaw in the WPA encryption protocol; the problem is the way manufacturers are implementing WPA in consumer products.

When you install WPA, you must create a passphrase that the software uses to generate a WPA key for encrypting and decrypting data. If you don't create a sufficiently complex passphrase, a cracker can recover the WPA key using software that compares the key against known dictionary words.

To prevent this, follow the same best practices for passphrase creation that you would when selecting a password. Use passphrases longer than 20 characters, don't use words found in the dictionary, and include random numbers and symbols.

Securing Clients and Hosts

While you should take the time to secure your access point with the techniques previously mentioned, you should also understand that your access point is still vulnerable to a skilled cracker. A cracker with advanced skills can circumvent many of the steps you have taken, and using network tools such as wireless sniffers and analyzers, can recover MAC addresses, IP addresses, and even encryption keys from network traffic.

Because of this, you should next turn your attention to securing the individual computers on your network. This way, even if a cracker manages to compromise your access point, he won't have access to the data on every computer connected to your network. This is not difficult to do and does not require you to have a great deal of networking skill.

Note If you have a hybrid network composed of wired Ethernet segments and Wi-Fi segments (see Figure 11-19), you must also secure the computers on the Ethernet segment of your network, particularly if you do not have a firewall separating the wireless segment from the rest of your network.

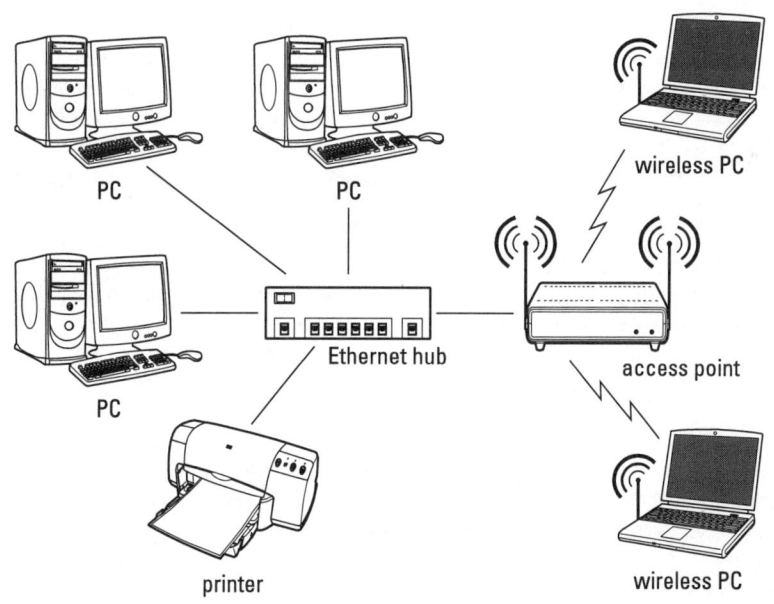

Figure 11-19: A hybrid network composed of Ethernet and Wi-Fi segments

Locking down Windows XP

The first step in securing individual computers on your network is to lock down the operating system. Provided here are steps that you can use to lock down Windows XP, including XP Pro and XP Home and Media Center, but they are applicable to Windows 2000 as well. I do not provide instructions for Windows 95/98/Me. Those versions of Windows are very insecure and easy to compromise. If you are using one of these older versions of Windows, consider upgrading to Windows XP immediately.

When securing Windows XP, the first three steps you should take are:

- Install and use a personal firewall.
- Get the latest Windows updates and patches.
- Install antivirus software and updates.

These are the three most important steps you can take, and following them will dramatically increase the security of your system.

Setting up personal firewalls

A personal firewall is a software application that resides on individual computers and monitors network connections. The firewall filters all incoming network traffic, blocking dangerous connections and unauthorized users. The better firewalls that are available also filter outgoing connections that originate from a machine. This is important because it can prevent malicious software like worms and Trojan horse programs from trying to spread to other machines on your network or to "phone home" to the hacker that sent them.

Windows XP comes with an Internet Connection Firewall (ICF) that you can enable to improve security on your machine. ICF is adequate for basic protection, but it doesn't filter or block outbound connections. ICF is a stateful firewall, which means it inspects the source and destination addresses on incoming data packets. ICF only allows incoming data through if it is in response to a request sent from your machine.

Unfortunately, this does nothing to stop malicious software such as a Trojan horse program from communicating with a cracker and receiving instructions. For this reason, I recommend that you purchase a third-party firewall like ZoneAlarm Pro from Zone Labs, or Symantec's Norton Personal Firewall that also filters outgoing connections.

Note Don't enable ICF on any connection that doesn't connect directly to the Internet. Because of the way it operates, ICF may interfere with local network connections between clients if you activate it on Ethernet cards or Wi-Fi adapters.

You can use third-party firewalls like ZoneAlarm to protect individual computers on a network. ZoneAlarm is highly configurable and can be set up so that it won't interfere with regular network functions.

If you do decide to use ICF, follow these steps to activate it:

STEPS: Activating ICF on Windows XP

1. **Left-click on the Start button.** The Start Menu appears.
2. **Under Pick a Category, click Network and Internet Connections.**
3. **In the Control Panel, click the Network Connections icon.** The Network Connections window opens (see Figure 11-20).

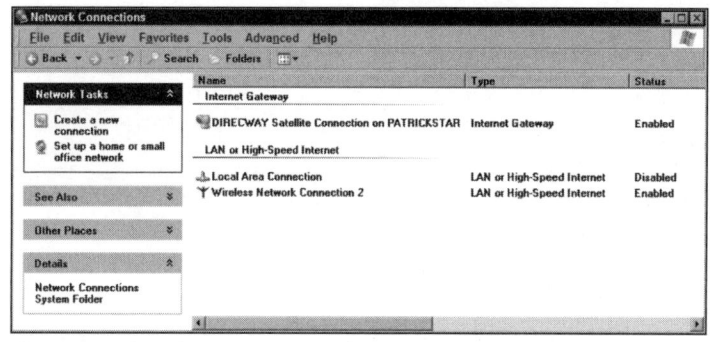

Figure 11-20: The Network Connections window

4. **Right-click your Internet connection, and select Properties from the menu that appears.** The connection's Properties dialog box opens.
5. **Click the Advanced tab (see Figure 11-21).**

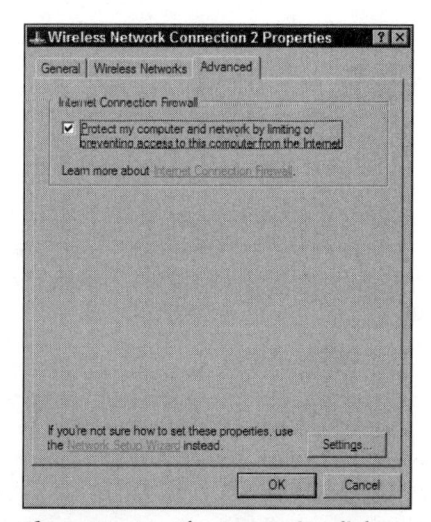

Figure 11-21: The Properties dialog box showing the Advanced tab

6. **Click the Protect my computer and network option.** The Settings button in the lower-right corner of the window is no longer grayed out.

7. **Click the Settings button and select the services in the Advanced Settings dialog box (see Figure 11-22) if you are running services like an FTP or Web server on your network and you want users to be able to access these from the Internet.**

8. **Click OK to close the windows and apply the settings.**

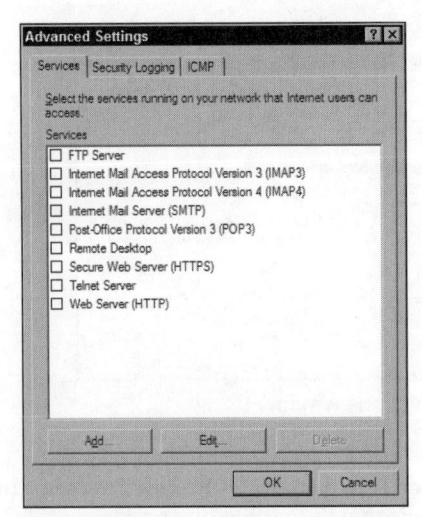

Figure 11-22: The Advanced Settings dialog box

Introducing the DMZ

Because wireless networks are more vulnerable than wired networks, a wireless segment connected to an Ethernet network may present a weak point that is vulnerable to attack. One method of reducing the risk to the wired segment of your network is to set up a demilitarized zone (DMZ) for the wireless connections.

A DMZ effectively isolates the Wi-Fi segment from the rest of your network by placing it outside a firewall (see Figure 11-23). The firewall between the WLAN and the LAN prevents unauthorized persons from connecting to the LAN if they compromise the access point or a wireless client.

Setting up a DMZ doesn't mean that you shouldn't still run personal firewalls on your WLAN clients. A DMZ does nothing to protect your wireless clients, just your wired ones.

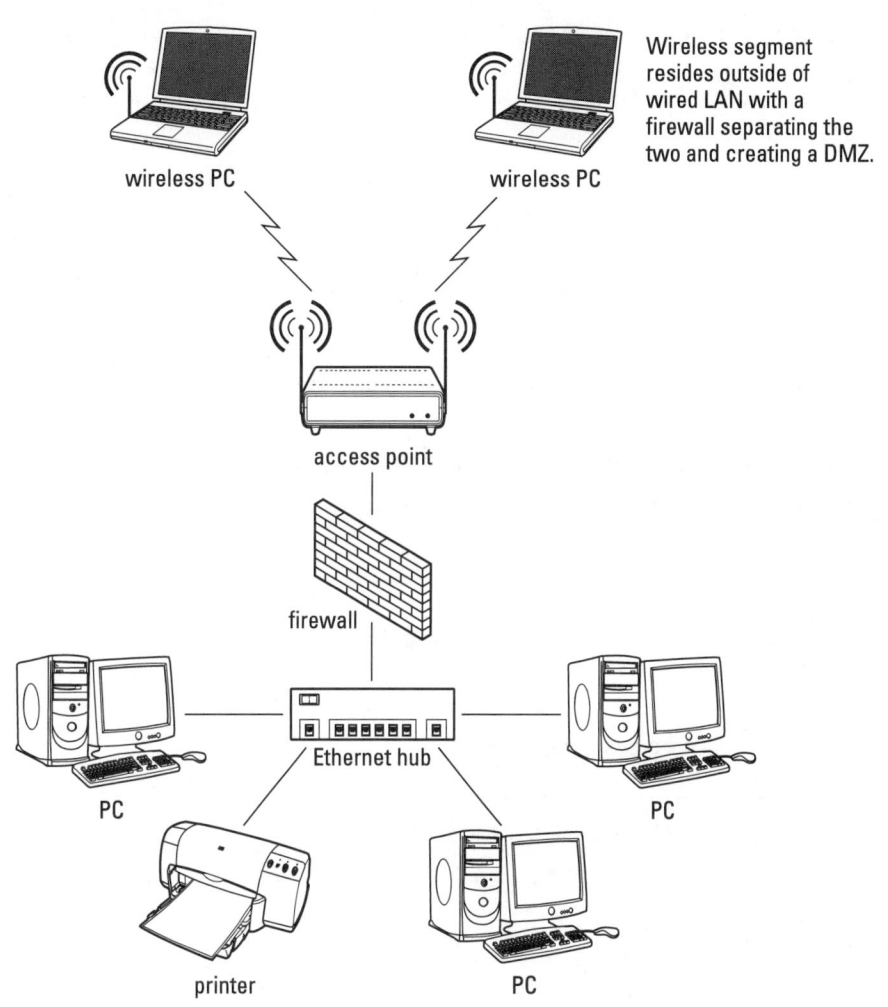

Figure 11-23: A WLAN setup in a DMZ

Get the latest Windows updates and patches

Researchers regularly discover new security vulnerabilities in operating systems and applications. Software bugs or oversight on the part of programmers causes these vulnerabilities. Occasionally the vulnerabilities arise due to unexpected results when different systems interact.

Crackers and virus writers regularly take advantage of known security flaws in the various Windows operating systems. They exploit these flaws to take control of computers or allow their viruses and worms to proliferate. MSBlaster, a worm that appeared in the summer of 2003, exploited a widely known security flaw in Windows operating systems.

Unfortunately, even though Microsoft had provided a patch to remedy the vulnerability weeks before the appearance of the worm, millions of users had not installed it. As a result, MSBlaster was able to spread quickly and cause millions of dollars in damage.

Software developers release patches and software updates to fix these vulnerabilities, and secure systems. Microsoft releases patches and updates for Windows XP and Windows 2000 on an almost weekly basis. It's important that you check for new updates, and download and install them to ensure that your system stays stable and secure.

To ensure that you keep Windows XP patched, you can automate the process by using Windows Update. Simply follow these steps:

STEPS: Keeping Windows XP patched using Windows Update

1. **Click the Start button.** The Start Menu opens.
2. **Click the Control Panel menu item.**
3. **Under Pick a Category, click Performance and Maintenance.**
4. **In the Control Panel, click the System icon.** The System Properties window opens (see Figure 11-24).

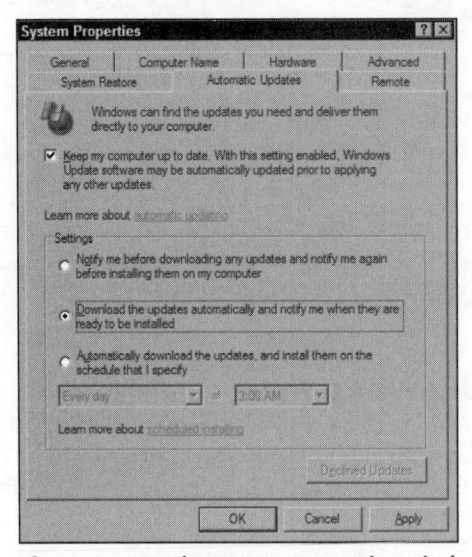

Figure 11-24: The System Properties window

5. **Click the Automatic Updates tab.**
6. **Click the Keep my computer up to date option.**

7. Click the **Download the updates automatically and notify me when they are ready to be installed** option.

8. Click **OK** to close the window and apply the new settings.

Windows Update will now automatically check for updates and download them when they are available and then notify you that they are ready to be installed. Alternately, you can set Windows Update to simply check for updates and then notify you when they are available. You can choose whichever method you are comfortable using.

> **On The Web**
>
> In addition to using Windows Update, I suggest that you periodically visit Security Focus (www.securityfocus.com). This is one of the most comprehensive security sites that you will find on the Web. Security Focus hosts a detailed vulnerability database pertaining to all operating systems and hardware devices that you can reference to help secure your own systems.

Install antivirus software and updates

With the number of worms, viruses, and other malicious programs that appear each month, it's imperative that you install a good antivirus program to protect your machine. There are many good commercial products available, and I recommend that you select one of them rather than using a freeware or shareware application.

Choose antivirus software that has an easy-to-use interface, and an efficient live update feature to keep the software current. This last feature is especially important because, like Windows, you need to update your antivirus software regularly to protect against new threats.

When selecting an antivirus program, consider the following points:

✦ Choose products from a developer with proven expertise and a good record of accomplishment. My personal selections (in order of preference) are Norton Antivirus, McAfee, and Panda antivirus.

✦ Choose a product with regular and *free* virus definition updates.

✦ Select an antivirus program that has an automatic or live-update feature so that it's easier for you to keep the software current.

✦ Choose a product that scans incoming and outgoing e-mail for infection.

Don't install and run more than one antivirus product at a time. This is definitely one case where more doesn't equal better; multiple antivirus applications can interfere with one another's operation, and may crash or slow your machine.

More steps to secure your XP machine

Once you have taken care of the three most important steps, you can consider implementing these remaining recommendations. They will help you lock down

Windows XP even tighter and further reduce the likelihood that your computer will be hacked.

- **Physically secure your machine.** Prevent unauthorized users from tampering with your computer, and never leave your laptop unattended in public.
- **Format all hard drive partitions with NTFS.** NTFS is more secure than the older FAT32 file system that older versions of Windows used (95/98/Me).
- **Use a router instead of Internet Connection Sharing (ICS) for shared Internet connections.** This is faster and will add another layer of security between your network and the Internet.
- **Use strong account passwords.** Refer to password best practices earlier in this chapter.
- **Disable unnecessary services.** See the following section for more information.
- **Disable the guest account.** You don't need it, so just turn it off
- **Use an account lockout policy.** You can set up Windows XP to lock your account after a certain number of failed attempts to connect. This helps prevent a cracker from trying hundreds of passwords to get into your account.

Disabling unnecessary services

Windows has many services enabled by default. Some of them are unnecessary, especially on stand-alone machines or machines on home networks. Leaving them running when they aren't used wastes computer resources at best, and at worst, they may be a security risk.

- **Computer Browser.** If you are on a stand-alone machine, you don't need this service. This service maps network clients and shares so that you can browse them. It only needs to be running on one machine for you to be able to browse the network.
- **Telnet.** Disable it. This is a terminal tool for connecting to remote computers, Ninety-nine percent of us will never use it, and it's a popular target for crackers.
- **Universal Plug and Play Device Host.** Not to be confused with Plug and Play (PnP), which is used for connecting devices such as video and sound cards to your machine (never disable PnP), UPnP is for connecting network devices. If you are using a Wi-Fi gaming adapter for a game console (Xbox, Gamecube, or PlayStation2) you may want to leave UPnP active if you encounter problems connecting with the game.
- **IIS.** This is normally not installed (not available on XP Home Edition), and I don't recommend that you do so unless you have to.
- **Netmeeting Remote Desktop Sharing.** If you don't use Netmeeting to hold meetings over the Internet, turn it off. If you use Netmeeting but don't need to share your desktop, turn it off.

- **Remote Desktop Help Session Manager.** This feature allows support personnel to connect to your computer. If you don't want or need remote help with your computer, turn this off.
- **Remote Registry.** Disable this if you don't need it. Making your registry available to remote users is a big risk. (This service isn't available on XP Home Edition.)
- **Routing & Remote Access.** This service is disabled by default in XP Home Edition. XP Pro users should also disable this service unless they need to dial into their computer remotely (via the modem).
- **SSDP Discovery Service.** This is used with UPnP. If you have UPnP disabled, turn this off as well.
- **Server Service.** If you are on a stand-alone machine or do not need to share files and printers from your machine, you can disable this.

If you disable a service and discover that an application or device needs it, you can always reactivate it.

Summary

In this chapter, I explained how your WLAN is vulnerable and presented steps that you should take to secure your WLAN. Among these were:

- Changing default passwords, SSID, and usernames
- Editing IP addresses
- Turning off DHCP and assigning static IP addresses
- MAC and IP address filtering
- Adjusting broadcast power and setting a minimum connection speed
- Using encryption
- Using a personal firewall
- Locking down client and host computers

Following these recommendations, you will be able to secure your WLAN so that it will be highly unlikely that a cracker or wardriver will be able to connect to your network.

✦ ✦ ✦

Protecting Your Wi-Fi Data

CHAPTER 12

♦ ♦ ♦ ♦

In This Chapter

Threats to wireless data

Selecting backup methods

Protecting hardware from damage

♦ ♦ ♦ ♦

In addition to securing your network and computers, there are other steps you can take to protect data on your WLAN. The best insurance policy against data destruction, whether malicious or accidental, is a good system of scheduled backups. This chapter discusses types of backups and ways to protect hardware.

Identifying Threats to Wireless Data

Chances are that you rely on your computer in some capacity, either at home or at work. Most of us have a great deal of accumulated data, including documents, music files, photos, and personal info. Have you ever stopped to think what you would do if that data were lost or destroyed?

I know I would lose time and productivity, and the loss would translate into financial loss as well while I worked to rebuild all of my records, and files. Fortunately, I back up my systems regularly, and I would be able to recover from data loss with minimal headaches.

There are many threats to your data; these include:

- ✦ Hardware Failure
- ✦ Software errors and application failure
- ✦ Media failure, such as hard drives
- ✦ Accidental damage to equipment
- ✦ Lost equipment, such as laptops
- ✦ User error and accidental deletion

Threats can originate from outside your network as well. These threats include:

- Malicious software, including viruses, worms, and Trojans
- Crackers and other computer criminals
- Natural disasters, such as earthquakes, fires, and floods
- Man-made disasters including terrorism, war, and industrial accidents

Of all of these threats, the most common reason for data loss is user error, followed by accidental damage, loss, and hardware failure. With the amount of attention paid to crackers and malicious software, you might think these would appear near the top of the list, but in reality, they're far less likely. However, while they aren't as likely they do a greater amount of damage, and you need to take them seriously and protect yourself with the techniques listed in Chapters 8 and 11.

In the next sections, I'll examine each of these threats in detail, and then begin discussing ways you can protect your data with backups.

Examining hardware failure

Newer hardware is more robust and less likely to fail than equipment developed even a few years ago. That said, these devices aren't indestructible and they do fail. Some of the causes for hardware failure include:

- Electrostatic discharge (ESD)
- Dust accumulation
- Excessive heat
- Power surges, usually due to lightening strikes
- Bad hard drive sectors
- Accidental damage, dropping, or spilling of liquids on hardware

I haven't presented these in any particular order of importance. Depending on the source, they're all touted as a major cause for hardware failure. Usually, vendors selling a solution to a particular problem will tend to rank it the highest. You should consider each of them, and protect yourself accordingly. The last section in this chapter deals with ways to protect your hardware from some of these threats.

Exposing outside threats to data

Threats from outside your network are the most publicized. As I said in Chapter 3, crackers make great headlines and get a disproportionate amount of media coverage compared to the damage that they actually do.

Malicious software also gets plenty of coverage, but you're more likely to have to deal with a virus or worm than you are a cracker. Viruses, Trojans, worms, and other computer pests are becoming all too common. If you fail to protect your computer using antivirus software, the likelihood of a virus or worm infecting it is very high.

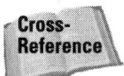

Cross-Reference Refer to Chapter 7 for information about protecting your computer against malicious software.

While I hope that you're unlikely to lose data due to a natural disaster, war, or terrorist attack, regular backups will help you to recover important data and get back on your feet should ones of these disasters occur

Choosing backup systems

The best protection against loss or destruction of your data is a good backup. Backing up data is the process of copying the data from the disk to another type of portable media, such as tape, external hard drive, CD, or DVD. Then, if a disk fails, you can recover and restore the data. The ability to perform Wi-Fi backups will be dependent upon your network's speed. Backing up large files, or even entire hard drives over a Wi-Fi connection can be time consuming. A better solution may be to create backup disks using removable media such as CDs or DVDs.

Determining what you need to back up

The first thing you must do is determine exactly what data you need to backup. You need to decide which data is most important to you, based upon how long it would take you to recover or recreate that data if it were lost. You should also consider the impact it would have on your life and your job if that data were gone tomorrow. With this in mind, consider backing up the following types of data.

- ✦ Personal information and data, including documents and e-mail
- ✦ Photos, video, and music files
- ✦ Internet settings, favorites, passwords, and settings for Web sites that you visit regularly
- ✦ Operating system and application software
- ✦ Customer data, databases, and financial data if you work from home

Some operating software can create a copy of itself on another disk, for restoration after hard drive failure or other damage. This usually only includes the operating system itself — not your personal files or applications you've installed. So, you should keep a backup copy of all of your software, in case of loss.

This list should help give you an idea of what you might need to back up. The cost of hard drives and removable storage media continually goes down. You should have no problem backing up all of your data at a reasonable cost.

Deciding how often you will back up

You've taken the first step by determining what you should back up; now all you have to do is determine how often that should occur. If you make frequent changes or regularly create lots of new data, like I do, then you'll need to backup frequently to keep your data safe. Of course, if you make infrequent changes, or produce little new data you can go longer between backups.

You can create *incremental backups*, which means backing up only those files that are new or that you've modified. This saves both storage space and time, and you can back up smaller amounts of data more frequently. Another type of backup you can conduct is a *differential backup*, which is useful when you only need the most current version of a file. Unlike an incremental backup, a differential backup doesn't save all versions of a file.

If you need to start from scratch, you can conduct a *bare metal backup*. This is a total restoration of your computer back to its original state. This is a common way for IT departments to get up and running after a problem, restoring the entire hard drive and starting over rather than trying to restore a portion and troubleshoot the problem.

Frequent backups will make it easier for you to get up and running quickly should you lose your data. A good backup schedule might include a full backup once a week and incremental backups every day. What data you have to back up and how often you will be conducting backups will help you choose a method of backing up and the type of media you will use.

Choosing a type of storage

You've determined what you need to back up and how often that needs to occur; now you can decide what sort of backup scheme is right for you. There are three general types of storage solutions; these are:

- Online storage
- Nearline storage
- Offline storage

Online storage

Online storage doesn't refer to backing up via the Internet. What I'm referring to here is a backup medium that's accessible in real-time, and doesn't require you to retrieve and load removable media such as recordable CDs (CD-R), or recordable DVDs (DVD-R). If you have more than one computer, you can have online storage on your WLAN, simply by backing up from one computer to another; if you have a dedicated computer for storing files, also called a file server, that's even better (see Figure 12-1). The goal is to keep your backups readily accessible to minimize the amount of time to restore your data.

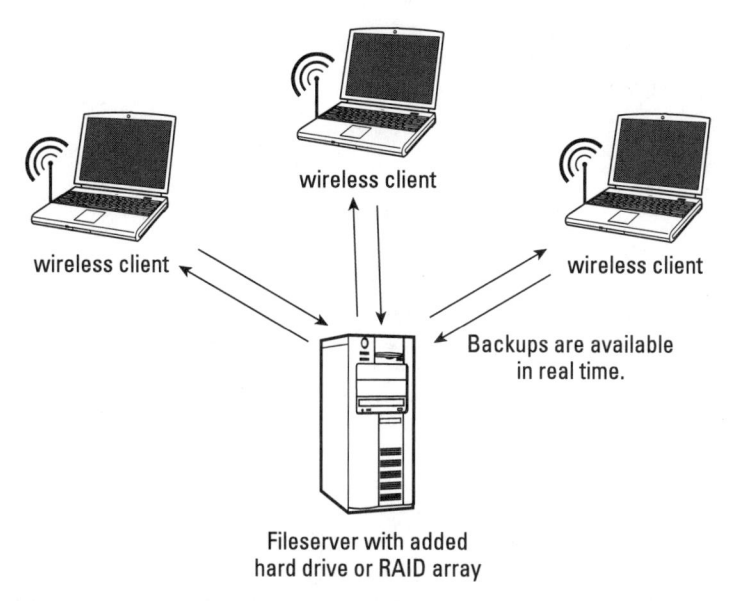

Figure 12-1: Online backups

> **Note** Backing up large files over a WLAN may take a considerable amount of time compared to backing up over an Ethernet connection. The speed of your WLAN will vary depending on which protocol you're using, connection strength, interference, and the number of users. Test your network by copying large files across it before you invest in a dedicated online storage system.

Ideally, online storage should offer redundancy by duplicating the data across more than one hard drive. If you have a dedicated PC set up as a server, you can achieve this using a redundant array of independent disks (RAID) array. Using a RAID array requires that you install a RAID hard drive controller, connected to two or more hard drives. The RAID controller writes data across multiple disks (see Figure 12-2). This process is called mirroring and creates redundant copies of your data. The hard drives in the RAID array usually appear as a single drive; if one drive should fail, your work will be uninterrupted as the remaining disks contain the same data.

There are different types of RAID systems available, including RAID 0, RAID 3, and RAID 5. A detailed discussion of RAID systems is beyond the scope of this book; just be aware that it's an option should you decide that your backup needs warrant it.

> **Note** If you are interested in more information about RAID systems I suggest you refer to *Highly Available Storage for Windows Servers* (Veritas Series) by Paul Massiglia (John Wiley & Sons, 2001).

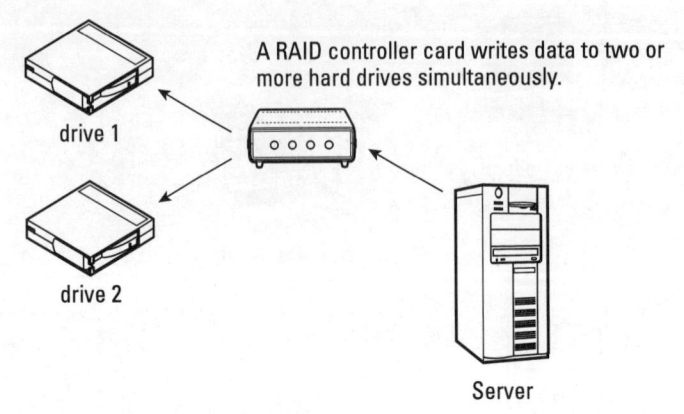

Figure 12-2: RAID controller

Nearline storage

Nearline storage is the route you are most likely to go when selecting a backup system. Nearline refers to creating backups on removable media and keeping those backups close by, rather than storing them away from your home (see Figure 12-3). You can create the backups using any sort of removable media including CD-R, DVD-R, and magnetic tape. Using nearline backups your data will be readily available should you need it. If you create duplicate copies of your backups, you can combine nearline storage with the next type of backup — offline storage.

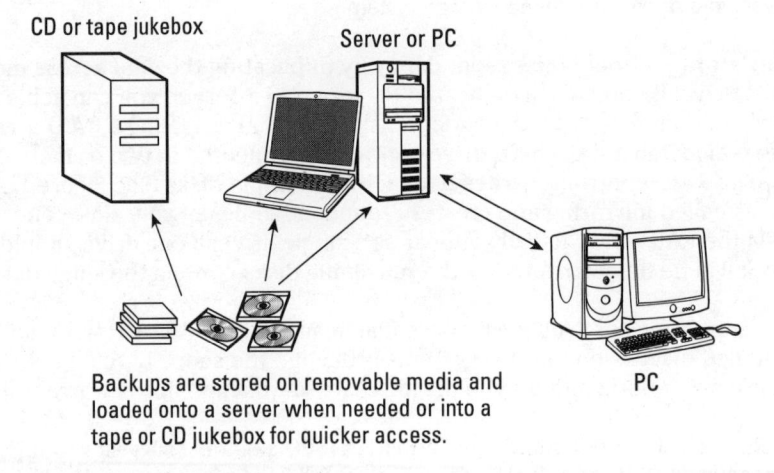

Figure 12-3: Nearline storage

Note: Some nearline storage solutions involve using CD-R, DVD-R or magnetic tape *jukeboxes*, devices that can hold multiple disks or tapes and load the appropriate backups as users request them. This sort of equipment would be overkill in a home or small office, as it's generally expensive and somewhat difficult to implement. However, if you have a file server with multiple CD-ROM or DVD drives you could create a scaled down version of this by loading your nearline backups into those drives and sharing them on the network.

Offline storage

Sometimes called offsite storage, offline storage involves creating your backups on removable media, and then storing those backups at another location. This is best done in conjunction with a nearline solution, as all this requires is that you make an extra copy of the data for offsite storage.

Offline storage offers the greatest protection for your data in case of catastrophe, such as a house fire or burglary. Having copies of your data stored at a relative's house or in a safe deposit box will ensure that you can recover from catastrophic loss of data.

Selecting media for backups

The type of storage system you decide to use will dictate what type of media you will be using. The type of media you select will affect the speed of your backups, and the speed at which you can retrieve your data. An online system using hard drives will often be the quickest, but will be more expensive than a nearline option that uses CD-R or DVD-R media.

I'm only going to present the media that you'll most likely use for a home WLAN; these include:

- Hard drives (internal, external, and removable)
- CD-R, DVD-R, and rewritable CD-RW and DVD-RW discs
- Magnetic tapes

Hard drives are fast, reliable, and relatively inexpensive hardware devices. You can install a hard drive within your computer or purchase an external hard drive that plugs into your computer via a serial, FireWire, or USB connection. Hard drives are ideal for online storage because of their speed; however, data on them is susceptible to corruption by disk error, user error, hardware failure, and malicious attack.

CD-R and DVD-R discs allow you to record or burn your data onto discs that you can easily store and transport. These discs can hold a significant amount of data, with a DVD-R being capable of storing up to 4.5GB of data. Individual discs are inexpensive; you can purchase CD-R discs for less than $.20 cents apiece and DVD-R discs for less than one dollar.

Once you've written data to CD-R and DVD-R discs, you cannot reuse the disc. However, there are rewritable CD-R and DVD-R discs available. These are known as CD-RW, and DVD-RW discs. They don't require a special drive to record, you can use any CD-R or DVD-R drive. However, some older CD-ROM drives may not be able to read them.

Magnetic tapes have been around for decades, originating with the giant reel-to-reel tape systems that have evolved into the small, compact cassette systems we use today. In my opinion, tape drives are an inefficient method of backup. Magnetic tape is sensitive and more prone to errors than other backup media, and because the data is stored sequentially along the length of the tape, it takes considerably longer to retrieve data than it does with other systems.

Tape backup systems have one advantage; they are usually automated so that you can schedule backups to occur automatically without your intervention. This is useful for daily backups as you can schedule these to occur during the night, when you're not using your computer.

Things you should consider

When selecting your backup method and media you need to consider four things:

- Cost
- Reliability
- Capacity
- Speed

I haven't presented these in order of importance, that's for you to decide. You may place greater emphasis on speed over capacity, or maybe reliability is your main criteria.

Over the last few years, the prices for drives and storage media have dropped significantly. Now you can purchase a DVD-R/CD-R combination drive for less than $150.00, and CD-R drives for less than $50.00. The cost of both internal and external hard drives has also dropped, with hard drives as big as 200GB selling for less than $120.00.

Tape drives continue to be the most expensive of these storage solutions, with drives that use 10GB tapes costing more than $670.00. The tapes usually cost more than $20.00 each. The price of these drives reflects the automation that they offer.

The most economical route for your home network is likely to be a CD-R/DVD-R combination drive. You can use this for nearline and offline storage, and combine it with online backup to another hard drive on your network.

The reliability of these devices varies. Hard drives are generally more reliable than tape drives, and even some CD-R and DVD-R media. However, hard drives are susceptible to mechanical failure. In this case, the data on the hard disk is usually unaffected and you can sometimes take it to a computer repair store and have them recover your data.

Tapes are most sensitive to environmental factors, and can degrade over time as well as be affected by electromagnetic fields (even weak ones). They are also susceptible to mechanical damage. On more than one occasion I've lost gigabytes of backed up data when a tape drive ate a tape.

While they are generally very reliable, the quality of CD-R and DVD-R discs varies depending on the manufacturer. Disc failure can result from physical damage, or by oxidation of the layer that contains the data. Personally, I've had the best luck with Sony and TDK discs. Most of these are still fine several years after I've burned them, while other brands have occasionally failed after only a year or two. CD-RW and DVD-RW generally have a higher failure rate than CD-R and DVD-R discs. Again, this varies, with some manufacturers producing better products than others.

Large hard drives and tape drives have excellent capacity for backing up large amounts of data. Both types of drive have capacities from 10 to over 200GB. Next come DVD-R discs, which have a capacity of 4.5GB, and CD-R discs, which can hold up to 800MB. Most backup systems that use DVD-R or CD-R discs allow you to write data to several discs if needed. Therefore, the storage limitations of DVD-R and CD-R should not be a problem for most home users.

If speed is your chief criteria, an online system using hard drives will probably be ideal. Hard drives write and read data quicker than other devices. If your WLAN is too slow when copying large files, you can back up to an external hard drive, and even move this drive from station to station to back up data from several network clients.

DVD-R and CD-R drives can read data quickly but take considerably longer to burn data. The slowest of all the devices are tapes drives, which take a long time to write data and even longer to retrieve it.

Backing up via the Internet

If you have a broadband internet connection you can perform quick backups of your important data via the Internet. This is advantageous for a few reasons:

- ✦ Your data is readily available like an online backup
- ✦ You can access this backup from any computer connected to the Internet
- ✦ It has the advantages of an offline backup, because your data is stored offsite

In my opinion, the best Internet backup and storage site is Ibackup.com. Ibackup supports Windows operating systems (95/98/ME/2000/XP), Mac OSX (for storage only), Pocket PC, and Palm OS devices.

To use Ibackup.com you have to register, choose a plan that's right for you, and download the appropriate client for your device and OS. Subscription start at $3.00 a month for 50MB of storage, and backup only plans start at $14.95 a month for 4.5GB. With subscription plans, you can share files, create sub accounts, and have ftp access.

> **On The Web**
> If you are interested in using Ibackup as an Internet storage solution visit www.Ibackup.com

Backup software

One of the most important components of any backup solution is the backup software or service itself. Although the primary function of a backup product is to create a copy of the data of your choice, many products have functions and features that can either enhance or detract from security. The intent here is not to compare products, but rather to help you understand what can enhance security relating to your backup process and the reliability of the restore process when needed.

It is a good practice to leverage encryption technologies like those used to protect sensitive data. Backup software should also contain some intelligence about what is being backed up. For example, there are backup software programs that understand that when an e-mail containing an attachment is sent to a group of users, only one copy of the attachment needs to be backed up. This can help to conserve media space while reducing the backup time.

> **Tip**
> You can go to www.storagesearch.com/backup.html for the most complete list of popular enterprise backup options.

Consider the following list of products when choosing options for your backup process:

- **Acronis True Image 7.0 (www.acronis.com).** This is a package that creates an exact disk image of your live system for complete backup. It currently sells for $49.

- **BackUp MyPC (www.stompinc.com).** A powerful, yet easy-to-use, data protection and disaster recovery solution that is designed for a single computer or a P2P network. The product backs up to recordable CD/DVD, tape, Zip, Jazz, and other removable media drives and employs disc spanning and data compression, which typically can double your media capacity for significant savings. It currently sells for $69.

- **BounceBack Professional 5.4 (www.cmsproducts.com).** Currently selling for $79, this set of backup and file management tools for users with an extra hard drive backs up data to a second drive, offering secure redundancy without requiring you to deal with removable media.

♦ **Drive Image 7 (www.symantec.com).** This is a powerful backup solution that allows you to protect all your valuable data. It currently sells for $69.95. As shown in Figure 12-4, you can easily back up everything on your computer: priceless digital photos, important financial records, your operating system, and all of your important programs and settings without ever leaving Windows. Save valuable time by scheduling automatic backups. You can quickly save your backups to virtually any writable CD or DVD drive and enjoy the flexibility of USB, FireWire, and Native Network support.

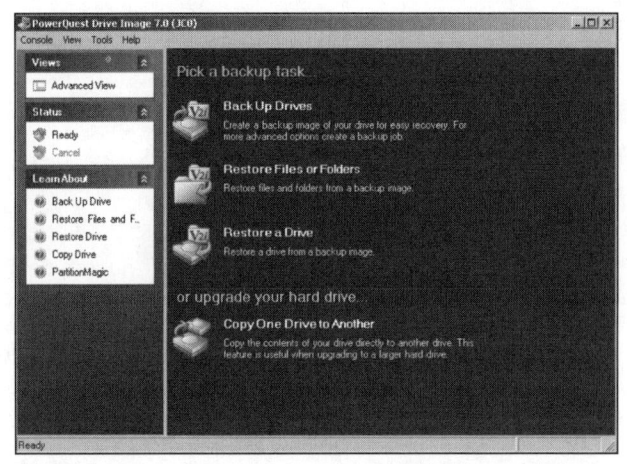

Figure 12-4: Drive Image is a powerful backup solution for home and small-office users with a simple user-friendly interface.

♦ **Ghost 2003 (www.symantec.com).** This award-winning software product (shown in Figure 12-5) protects your data from computer disasters. The intuitive Windows interface makes it easy to create regular backups of your hard drive, and you don't even need to make a boot disk. It currently sells for $69.95.

♦ **Iomega Automatic Backup (www.iomega.com).** This software suite backs up your important files automatically. Simply select the files you want to protect and personalize your backup schedule and location(s). Restoring data is just as easy — all you do is drag and drop. It's currently priced at $39.95.

♦ **Norton GoBack 3.0 (www.symantec.com/goback/).** You can roll back your system to a healthy state after a system problem or user error, retrieve deleted files easily, and recover individual files or an entire hard drive using this software. It currently sells for $39.95.

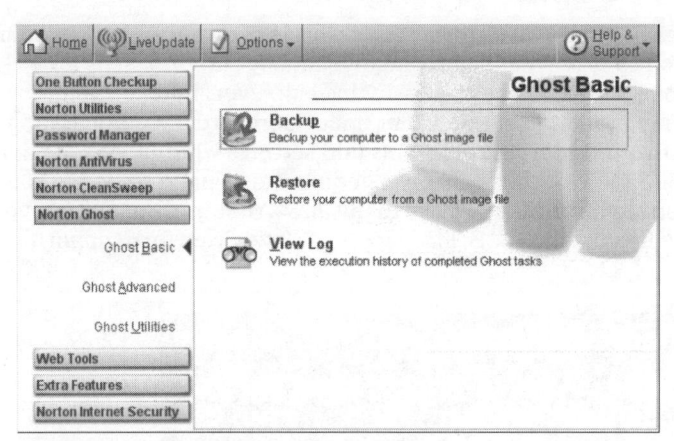

Figure 12-5: Ghost's simple interface will get you underway quickly.

✦ **Retrospect Professional 6.5 (www.dantz.com).** This product is designed for home and small-office users who want to protect a single Windows computer and up to two additional networked Windows or Macintosh desktops and notebooks by backing them up to a Windows-based computer. It's currently priced at $89.

Protecting Your Hardware

What is another way to protect your Wi-Fi data? Protect your WLAN assets, such as antennas and access points, from common threats. These threats include damage from electrical surges and failing power supplies.

Surge and spike protection

Power surges and spikes cause the most common forms of electrical damage to systems. In effect, surges and spikes cause an overflow of power to your circuit that can damage your equipment unless you take proper protective measures. Surges and spikes can originate from lightning, high-power electrical devices — including your air conditioner and refrigerator — and things such as faulty wiring and problems with the power lines or electric company equipment.

There are many types of devices available that can protect your equipment from most types of surges and spikes. Some of them even include protection and noise filtering for phone, coaxial, and network cable connections. Most types of protective hardware have limitations though, so be sure to choose the product with a protection level that you are comfortable with.

To protect your equipment, you should have individual surge protectors for each outlet used with sensitive equipment. The following list covers some of the choices you have to consider:

- **Basic power strip.** This is usually an extension with extra outlets and some basic protection. Basic strips are typically inexpensive and should really only be used for low-end, low-power, inexpensive equipment such as a clock radio.

- **Advanced power strip/station.** For anywhere from $12 to $30 you can get a power strip surge protector with better ratings and extra protection. These usually offer features such as network or phone and cable protection and some type of insurance against loss due to failure to protect your devices. Be sure to check the ratings on these devices, and use them accordingly. For example, you may only be covered for a computer and monitor on a single power outlet.

- **Uninterruptable Power Supply (UPS).** A UPS is designed to offer continuous power during an outage — just enough to manually power down your devices safely. These cost anywhere from $35 to over $150 and usually protect against most forms of electrical issues in accordance to ratings. A UPS is the recommended protective device for your wireless access points and WLAN hardware. Be sure to choose one that supports the power requirements for your hardware.

Overall, be sure to check the Underwriters Laboratories (UL) ratings for your power protection hardware. UL is an independent company that tests electric and electronic products against standard safety criteria. On any given surge protector, for example, you should find ratings such as clamping voltage (the lower the better), energy absorption/dissipation (the higher the better), and response time (less than one nanosecond is recommended).

With regard to your WLAN access points, to avoid interference to its services, they should be physically located away from external sources of electromagnetic interference, such as microwave ovens. In addition, if you install your access point outdoors it should be in a waterproof enclosure.

Lightning arrestors and grounding antennas

For added protection for your wireless antennas, you should consider deploying what is called a lightning arrestor. This device is used to protect your equipment from nearby lightning strikes. It is typically an inline device with a grounding ring that connects to the cable between your antenna and receiver. See Figure 12-6 for an example diagram of a lightning arrestor.

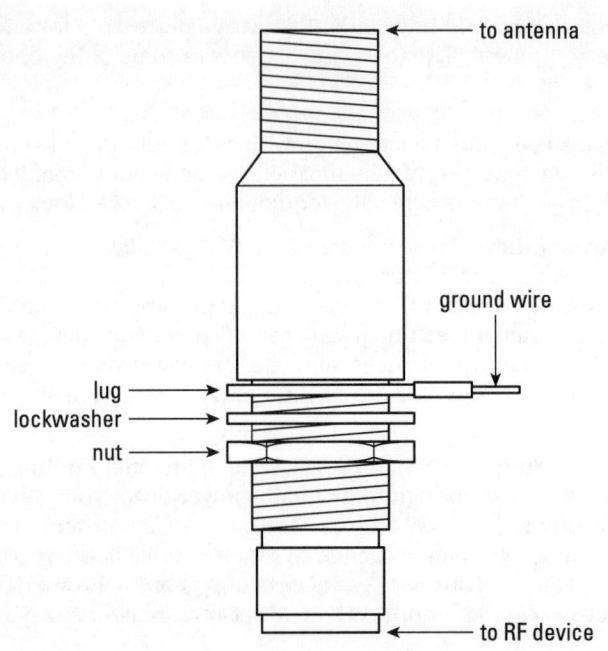

Figure 12-6: A lightning arrestor helps prevent damage due to lightning-induced surges or static electricity.

You need to connect your lightning arrestor to an earth ground, and when doing so it should protect your devices against spikes up to 5,000 amps. Be sure to follow the grounding directions to the letter, and be prepared, as the direcitons may say to connect it to the building's structural steel, a building reinforcing bar, a standing water pipe or "fire riser," an earth down-conductor, a lightning rod, or an 8-foot copper or steel rod placed into the ground.

This is a great first-line-of-defense mechanism against damaging electrical storms to deploy if your budget permits it. Depending on the unit type, it can cost anywhere from $40 to several hundred dollars.

Cross-Reference: For more information about grounding antennas to avoid static buildup and interference refer to Chapter 8.

Preventing problems through proper maintenance

Another way you can protect your hardware and prevent failure is through periodic cleaning and maintenance. You will need to clean your PCs inside and out, and keep all of your WLAN hardware free of dust.

Note: Never open up any of your WLAN devices, laptops, or computer monitors. The only hardware you can open for cleaning is your PC. Even then, be sure that opening the case does not void any of your warranties. If this is the case you can take your PC to any computer store with a service department. They will be able to clean your PC for you.

To clean and maintain your PC you will need the following equipment:

+ A grounding wrist strap
+ A grounding mat, for added protection
+ A computer vacuum, rated for use inside a PC
+ Compressed air, usually sold in cans
+ A Philips screwdriver

Caution: Always, disconnect the power to your PC *before* you open the case. By disconnect, I mean unplug the computer from the outlet, don't just turn it off. Also disconnect the monitor and other peripherals from the computer prior to opening the case. I've met people that have been told that it's safer to work with the unit connected to an outlet. This is dangerous misinformation, and you could get injured or killed if you attempt it.

Because static electricity can destroy circuits, it's important that you discharge any built up static charge from your body prior to beginning this work. Do this by touching a metal surface.

Next, you should put on the antistatic wrist strap and clip one end of it to bare metal inside the computer case. Don't clip the strap to any circuit boards, chips, or drives as doing so can damage them. Connect only to bare metal within the case. The wrist strap grounds you to the PC case and maintains and equal charge between your body and the PC, avoiding any static sparks that could damage components.

Before using the compressed air, read the directions carefully. You should spray compressed air in short bursts and at a minimum distance of 6 to 8 inches from components. Prolonged spraying can freeze components or cause condensation to appear on them that could destroy them. With the compressed air, you can blow any dust off components, out of vents, and out of the power supply and drive bays.

You can now use a micro-vacuum rated for use inside computers. Using a regular vacuum may ruin your PC. Micro-vacuums are designed to be safe for use in computers and don't generate static or conduct electricity. Don't allow the vacuum to contact components — use it at a safe distance of 3 to 4 inches.

Pay special attention to cleaning out the power supply, vents, and the heat sink on your PCs processor. Dust build up in these areas can cause the components to overheat, or even short-circuit.

Summary

In this chapter, I presented different ways to protect your Wi-Fi data by implementing a backup and recovery plan and protecting your hardware from electrical anomalies. These topics included:

- Threats to your Wi-Fi data
- Determining what you should back up
- How often you should back up
- Choosing a storage method
- Selecting media for backups
- Important things to consider, including speed, capacity, and reliability
- Backing up via the Internet
- Different backup software
- Protecting your hardware from power surges and spikes
- Proper maintenance

It is a good practice for any home user to follow a regular backup schedule. And, you should also deploy the appropriate power protection with an optional lightning arrestor and an approved grounding system to protect your systems from electrical threats.

✦ ✦ ✦

Suppliers, Manufacturers, and Resources

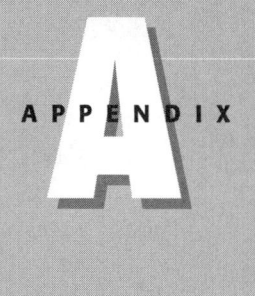

Wi-Fi Networking Hardware

The following manufacturers produce wireless hardware targeted toward home, small office, and home office consumers, in addition to other networking and computing products and accessories.

3COM (www.3com.com)

Founded in 1979, 3COM manufactures many networking devices including Wi-Fi gear. The 3 "COMs" in the name are *computer, communication,* and *compatibility.*

Wireless Products: access points, adapters, antennas, Bluetooth adapters, bridges, and routers

Actiontec (www.actiontec.com)

Actiontec produces broadband and wireless networking products for consumers. Actiontec focuses on producing affordable products for home users that are easy to use.

Wireless Products: access points, adapters, Bluetooth adapters, and DSL router/modems

AmbiCom (www.ambicom.com)

Founded in 1997, AmbiCom produces a full line of wired and wireless products. AmbiCom specializes in mobile wireless solutions and produces wireless adapters for laptops, PDAs, and printers. In addition to wireless products, Ambicom also produces GPS cards for Pocket PC devices.

Wireless Products: access points, adapters, Bluetooth adapters, GPS cards, and routers

Apple Computer (www.apple.com)

Steve Wozniak and Steve Jobs founded Apple in 1976. Since then, Apple has been one of the most innovative companies in the computer industry. Although known for its operating systems and Computers, Apple also manufactures wireless products for both Mac OS and PC systems.

Wireless Products: AirPort Extreme wireless access point

Belkin (www.belkin.com)

Belkin has been producing networking products since its founding in 1983. Now it has expanded its product line to include wireless networking products as well.

Products: access points, adapters, antennas, Bluetooth adapters, and routers

Buffalo Technology (USA) (www.buffalotech.com)

Buffalo Technology (USA) is a subsidiary of Japan's Buffalo, Inc. Buffalo manufactures wired and wireless networking devices in addition to its computer memory products.

Products: access points, adapters, antennas, bridges, repeaters, and routers

D-Link Systems (www.dlink.com)

D-Link produces wired and wireless networking products for the home, small office and enterprise environments. It also produces many other wireless products, including cameras, and wireless game adapters. D-Link is one of leaders in products for home use and has excellent technical support.

Products: access points, adapters, antennas, Bluetooth adapters, bridges, cameras, repeaters, and routers

Hawking Technology (www.hawkingtech.com)

Hawking Technology has been producing wired and wireless networking products for over 13 years. Hawking specializes in products for home and small business users and has excellent customer support.

Products: access points, adapters, antennas, Bluetooth adapters, print servers, and routers

Linksys (www.linksys.com)

Linksys is a division of Cisco Systems that focuses on consumer networking products. Linksys manufactures consumer wired and wireless networking products.

Products: access points, adapters, antennas, cameras, game adapters, Bluetooth adapters, print servers, wireless media centers, and routers

MiLAN Technology (www.milan.com)

Founded in 1990, MiLAN manufactures wired and wireless networking gear for home and business. MiLAN's focus is on business products, but it does produce some excellent and affordable Wi-Fi gear suitable for home use.

Products: access points, adapters, and bridges

NETGEAR (www.netgear.com)

Founded in 1996, NETGEAR focuses entirely on producing wired and wireless networking equipment for home and small-office users.

Products: access points, adapters, antennas, bridges, print servers, and routers

SanDisk (www.sandisk.com)

Founded in 1988, SanDisk is the largest producer of flash memory storage in the world. SanDisk also produces some Wi-Fi cards for Pocket PC devices.

Products: CF card Wi-Fi adapters and flash storage.

SMC Networks (www.smc.com)

Founded in 1971, SMC has been an industry leader for over 30 years. SMC produces a full line of consumer wired and wireless networking products, including the EZ Networking line specifically designed for home and small-office users.

Products: access points, adapters, bridges, and routers.

Startech.com (www.startech.com)

Startech manufactures a variety of replacement parts and computing and networking hardware, including wireless gear. Startech specializes in hard-to-find parts and has excellent support and service.

Products: access points, Bluetooth adapters, adapters, bridges, and routers

TRENDware International (www.trendnet.com/us.htm)

Founded in 1990, TRENDware is a leading manufacturer of networking products worldwide. TRENDware manufactures both Ethernet and wireless networking products for home and small business users.

Products: access points, adapters, antennas, Bluetooth adapters, bridges, print servers, and routers

U.S. Robotics (www.usr.com)

Founded in 1976, U.S. Robotics is one of the world's leading manufacturers of networking devices, including wireless gear. Chances are that at some point, you've connected to the Internet with a USR modem. USR produces devices for all markets, from home consumers through the enterprise.

Products: access points, adapters, bridges, and routers

Wi-Fi Resources

The following Web sites are general hobbyist or user group sites with good information about Wi-Fi:

Bay Area WUG (www.bawug.org)
The homepage of the Bay Area (San Francisco) wireless users group

Dallas/Ft. Worth WUG (www.dfwwireless.org)
The homepage of the Dallas/Ft. Worth wireless users group

Dr. Trevor Marshall's Web site (www.trevormarshall.com)
The homepage of Dr. Trevor Marshall, this is a great site with information about Wi-Fi security, and antenna designs. A good place to start if you want to try building your own antenna.

Jason Hecker's Wireless Goodies (www.wireless.org.au/~jhecker)
This is a good site with plans for a 2.4 GHz helical antenna, an excellent place to visit if you want to try building your own helical antenna.

NYC Wireless (www.nycwireless.net)
The homepage of the New York City wireless users group

The Homebrew Antenna Shootout (www.turnpoint.net/wireless/has.html)
A great page with lots of good antenna ideas, an excellent starting place if you're interested in building an antenna

San Diego WUG (www.sdwug.org)
The homepage of the San Diego wireless users group

Seattle Wireless (www.seattlewireless.net)
A Seattle WUG with the goal of creating a free wireless community network

So Cal WUG (www.socalwug.org)

The homepage of the Southern California wireless users group (Los Angeles area)

Wireless Anarchy (http://wirelessanarchy.com)

A good wireless portal with links to several wireless user groups

Security, Privacy, and Antivirus Resources

I like to stress security, and I'm going to stay in character here by recommending some good sites for security, privacy, and antivirus information.

CERT (www.cert.org)

Located at the Software Engineering Institute (SEI) at Carnegie Mellon University, the Computer Emergency Response Team (CERT) is a reporting house for security incidents and vulnerabilities. CERT is a great place to get facts about new threats or for organizations to report serious security incidents.

CIAC Hoax Busters (http://hoaxbusters.ciac.org)

The U.S. Department of Energy Computer Incident Advisory Capability (CIAC) maintains an informational site about e-mail hoaxes. Hoaxes cause many problems, especially when thousands of people forward them to friends, coworkers, and relatives. Next time you get a virus warning, or great moneymaking offer, check here before you forward the e-mail. You'll save your friends time, and you won't contribute to the flood of useless e-mail.

Electronic Frontier Foundation (www.eff.org)

The EFF is a nonprofit organization committed to protecting our right to "think, speak, and share our ideas, thoughts, and needs using new technologies, such as the Internet and the World Wide Web." The EFF works to protect free speech and privacy rights for all people, visit and support the EFF.

Shields Up!! Gibson Research Corporation (www.grc.com)

The Shields Up!! Internet Security Checkup helps you determine how vulnerable your computer is and tests your firewall software. Created by *InfoWorld* columnist and programmer Steve Gibson, millions of people have taken advantage of this free tool, and you should, too. You may be surprised at the results you'll see.

Security Focus (www.securityfocus.com)

Security Focus maintains the Security Focus vulnerability database and the Bugtraq security mailing list. Security Focus is one of the most comprehensive security sites on the Internet, you should check this site often to stay informed of vulnerabilities that affect your computer and other devices.

Spybot Search & Destroy (www.safer-networking.org)

This site has useful software that can detect and remove spyware from your computer. Spyware is software that tracks your surfing habits and reports your private information to advertisers. Spyware also includes malicious applications that record everything you type on your keyboard (including passwords), for someone else to read.

Warchalking (www.warchalking.org)

Warchalking.org is an informational site about the practice of warchalking. It includes a sample of warchalking symbols and the history of warchalking.

Wardrive.net (www.wardrive.net)

Wardrive.net is a great informational site with links to wireless security resources. This is a great place to start, whether you're securing your WLAN or doing further research.

World Wide WarDrive (www.worldwidewardrive.org)

This is the official site of the World Wide WarDrive. The World Wide WarDrive was organized to generate awareness of the security issues around Wi-Fi access points and to compile statistical data. Participating members include hobbyists and professionals.

www.vmyths.com (www.vmyths.com)

Vmyths.com exists solely for the "eradication of computer virus hysteria" and is run by (actual) virus expert Rob Rosenberger. Vmyths.com provides straight talk about virus threats, dispels myths, and exposes virus hoaxes. Visit this site and get the facts about computer viruses from experts that aren't affiliated with any anti-virus company.

PDA, PC, and Accessory Manufacturers

Here's a list of the major manufacturers of popular PDAs, PCs, and accessories. It's not exhaustive by any means, but it's a good start.

Dell (www.dell.com)

Founded in 1984, DELL is one of the largest personal computer manufacturers in the world; Dell also produces a line of handheld Pocket PCs.

Garmin (www.garmin.com)

Founded in 1989, Garmin primarily manufactures handheld GPS units (which can be used when wardriving) and produces the IQue, the first Palm OS PDA with integrated GPS capability.

Gateway (www.gateway.com)

Ted Waitt founded Gateway in 1985 in his Iowa farmhouse. Since then Gateway has grown into one of the world's largest personal computer manufacturers. Gateway focuses on providing consumers with custom PCs and consumer electronics.

HP (www.hp.com)

Founded in 1939 by Bill Hewlett and Dave Packard, HP is one of the world's largest manufacturers of digital equipment, including PCs, printers, and networking gear. HP also produces the popular iPAQ series of Pocket PC handhelds.

Kensington (www.kensington.com)

Founded in 1981, Kensington is a leading manufacturer of peripheral equipment for computers and PDAs. In addition to peripherals, Kensington also produces the Wi-Finder, a consumer device for locating Wi-Fi hotspots.

palmOne (www.palmone.com)

Palm Computing was founded in 1992, acquired by U.S. Robotics (and then 3Com), and spun off as an independent company again in 2000. In 2003 after acquiring Handspring, Palm became palmOne. Today palmOne manufactures Palm OS handhelds.

StuffBak (www.stuffbak.com)

An innovator in property identification, StuffBak produces ID labels for all of your personal property, including computing equipment. Through its unique program, StuffBak facilitates the return of your lost property. Because the labels really work, I highly recommend that you use StuffBak's identification labels on all your mobile devices. The service is inexpensive, and easy to use.

Software

The following is a list of software of particular interest to wireless users, including antivirus, firewalls, and sniffer utilities. Although I've only included information for Windows applications, some of the sites include products for other operating systems.

ApSniff (www.bretmounet.com/ApSniff)

ApSniff is a wireless access point sniffer that runs on Windows 2000 systems.

APTools (http://winfingerprint.sourceforge.net/aptools.php)

APTools is an excellent sniffing tool for detecting rogue access points. APTools is available for Windows 2000 or Unix systems.

Internet Security Systems (http://blackice.iss.net)

ISS is the developer of the personal firewall BlackIce, available for Windows OS systems.

McAfee (www.mcafee.com)

McAfee is the developer of antivirus and security-related applications, including VirusScan, and Personal Firewall Plus.

Netstumbler (www.netstumbler.com)

Network Stumbler is the most popular stumbler application for Windows 2000/XP. The site also hosts Mini-Stumbler, a stumbler application for Pocket PC devices.

Symantec (www.symantec.com)

Founded in 1982, Symantec is the leader in antivirus and information security software. Symantec produces the Norton brand of consumer security products, including Norton Anti-Virus, Norton Personal Firewall, and other utilities.

ZoneLabs (www.zonelabs.com)

ZoneLabs produces ZoneAlarm, one of the easiest-to-use and most effective personal firewalls available. If you don't have an integrated firewall on your router or access point, I recommend that you install this product.

◆ ◆ ◆

APPENDIX B

Wireless Standards

There are more wireless standards than just the big three (802.11a, b, and g). Here is a more complete list of the standards that relate to WLANs. You don't need to memorize any of these, although you will be able to amaze your friends, coworkers, and the guy at the computer store by dropping a few of these names when conversations turn to wireless.

802.11b

Also known by the consumer-friendly name Wi-Fi (which also includes 802.11a and 802.11g) 802.11b arrived in 1999 and rose to dominate the home, small office, and home office markets. In the process 802.11b sent the competing HomeRF standard to an early grave.

Because it uses the unregulated 2.4 GHz radio band, consumers do not have to worry about licensing to use 802.11b gear, and due to its market dominance, there is an abundance of inexpensive 802.11b equipment available. 802.11b shares the 2.4 GHz band with other electronic devices (such as cordless phones) and these can sometimes interfere with its normal operation. The maximum data rate of 802.11b devices is 11 Mbps, although actual throughput seldom exceeds 6 Mbps.

802.11a

Unlike 802.11b and 802.11g, 802.11a operates in the 5 GHz radio band, instead of the 2.4 GHz band. Because of this, 802.11a is not compatible with 802.11b or 802.11g devices, and upgrading your WLAN to 802.11a will require you to purchase new equipment. However, you can purchase dual-band access points that use both radio frequencies and can operate on both 802.11b and 802.11a WLANs.

Although 802.11a has a maximum data rate of 54 Mbps, actual throughput seldom exceeds 35 Mbps, and is often less than 20 Mbps. Usually, 802.11a can only achieve its top connection speed if the distance between adapter and access point is less than 30 feet. Because of its short range, 802.11a may not be your best choice if you have to cover a wide area with Wi-Fi access. Doing so will require you to buy more access points than another standard might require (specifically 802.11g).

802.11g

The newest Wi-Fi standard, 802.11g is an extension to 802.11b and provides the throughput of 802.11a while using the 2.4 GHz band. 802.11g is backward compatible with 802.11b and can coexist on the same WLAN. However, mixed standard WLANs (802.11b and 802.11g) often experience slower throughput speed than WLANs implementing a single standard. 802.11g is susceptible to the same sources of interference as is 802.11b because it also operates in 2.4 GHz band.

802.11c

The 802.11c standard isn't relevant to Wi-Fi installers or end users. This standard addresses bridge operations, and engineers designing access points use it. Just think, now you can amaze your friends with this pearl of Wi-Fi wisdom.

802.11d

Not all countries make the 2.4 GHz and 5 GHz bands available for unlicensed use like the United States does. The 802.11d standard exists to make 802.11 networking acceptable to countries using different frequency bands. Once again, no need to memorize this; it's just useful geek trivia.

802.11e

The 802.11e standard is a quality of service (QoS) extension to the MAC for prioritizing traffic within the 802.11 standard in order to provide services for data, voice, and video. Being a MAC extension, 802.11e is common to all of the physical layer standards (802.11b, 802.11a, and 802.11g).

802.11f

802.11f is an addition to the base 802.11 protocol that facilitates interoperability between access points from different vendors within a multivendor network. Eventually, 802.11f will give you the freedom to roam within a multivendor Wi-Fi network without losing your connection.

802.11h

The *h* in 802.11h stands for HiperLan, a competing WLAN standard in Europe. In many European countries, radar and satellite services operate near the 5 GHz band, and without modification 802.11a systems could possibly interfere with them. 802.11h is a supplement to the 802.11 MAC layer, specifically for 802.11a, so that it can operate in the 5 GHz band in Europe in compliance with EU regulations. The addition of 802.11h should facilitate adoption of 802.11a by European users.

802.11i

Wired Equivalent Privacy (WEP), the encryption algorithm implemented in current WLANs, is severely flawed and crackers can defeat it easily. In response, the IEEE is developing 802.11i as a new supplement to the 802.11 MAC standards in order to overcome the security weaknesses in Wi-Fi. 802.11i improves encryption, and the new Wi-Fi Alliance developed Wi-Fi Protected Access (WPA) encryption based upon a portion of the 802.11i standard. The next iteration of WPA, WPA2 will be fully compliant on the published 802.11i standard.

802.11 IR

The IEEE developed the 802.11 IR wireless standard around the same time as 802.11. Rather than using radio waves, 802.11 IR uses ultrahigh frequency infrared light (beyond that which the human eye can see) to send data. 802.11 IR operates at the same speeds as the original 802.11 specifications.

No manufacturer that I know of has used this standard in consumer products. There are some IR networking devices that utilize proprietary standards that are close to 802.11 IR, however these exist in niche markets (industrial controllers, security systems, and some education networks).

If developed commercially, 802.11 IR could be useful in high-security networks. Because infrared light doesn't go through walls, it is almost impossible to intercept or interfere with the signal from outside a facility (as opposed to radio waves, which propagate beyond a building's walls where a third party can intercept them).

802.11n

The IEEE has created a working group to begin investigating the next generation of high-speed wireless standards. One proposed standard, 802.11n, is approximately three years away but will possibly provide more than 100 Mbps of actual throughput (not data rate, throughput) for WLANs.

802.16a and WiMAX

802.16a is a standard for wireless metropolitan area networks (WMAN), and utilizes the 2–11 GHz frequency band at speeds up to 280 Mbps. 802.16a will have a maximum operating range of up to 30 miles. 802.16a will provide wireless broadband to portable and fixed users and connect Wi-Fi hotspots to the Internet. WiMAX is an organization to promote 802.16a, similar to the relationship between the Wi-Fi Alliance and the 802.11 family of standards.

Bluetooth and 802.15.1

Bluetooth is a wireless personal area networking (WPAN) technology that operates in the 2.4 GHz frequency band. Developed by the Bluetooth Special Interest Group (founded by Nokia, Ericsson, IBM, Intel, and Toshiba), Bluetooth is named for the Danish King Harald Blåtand (Bluetooth) who unified Denmark and Norway in the tenth century. Like King Bluetooth, the Bluetooth standard intends to *unify*, in this case the telecom and computing industries.

Bluetooth is a complementary standard to 802.11x, Bluetooth devices peacefully coexist and in some cases interoperate with Wi-Fi devices. Bluetooth doesn't compete with Wi-Fi standards; its range is too short and its throughput speed is too slow (1 Mbps).

Bluetooth allows users to connect many different computing and communications devices easily and simply without cables. Wi-Fi devices replace Ethernet cables, and Bluetooth connects peripherals without all those annoying cables. Keyboards, optical mice, printers, digital cameras, and PDAs that employ Bluetooth are already available.

802.15.1 standard is fully compatible with the existing Bluetooth spec. IEEE created 802.15.1 around a licensed portion of the Bluetooth specification. As Bluetooth develops further, the IEEE will likely incorporate these changes into the 802.15.1 family of standards.

802.15.4 and ZigBee

The 802.15.4 standard is a low-speed WPAN technology that operates in the 868/915 MHz and the 2.4 GHz bands. ZigBee is an alliance of corporations seeking to provide the applications and networking layer for the 802.15.4 MAC. ZigBee has low power consumption and low data rates and is secure and reliable. This is a promising standard for remote monitoring and control applications where devices don't need to send large amounts of data. ZigBee is named for the zigzagging flight of bees, which, like the nodes on a 802.15.4 network, are simple individually but can accomplish complex tasks when combined.

802.15.3 and WiMedia

802.15.3 is a WPAN standard that operates in the 2.4 GHz frequency and can coexist with 802.11b without incident. 802.15.3 supports speeds up to 55 Mbps and purportedly will connect up to 245 wireless devices in an ad hoc fashion. These devices will be able to sense and connect to the network without facilitation from a user. Somewhat analogous to the Wi-Fi Alliance, the WiMedia alliance is promoting 802.15.3 and hopes to do for 802.15.3 what Wi-Fi has done for 802.11x.

HiperLAN

Developed by the European Telecommunications Standards Institute (ETSI), *High-Performance Radio Local Area Network* (HiperLAN) is the wireless networking standard throughout most of Europe. There are actually two HiperLAN standards—HiperLAN/1 and HiperLAN/2—and they operate at speeds up to 20 Mbps and 54 Mbps, respectively. Like 802.11a, both standards operate in the 5 GHz frequency band.

Because it is compatible with 3G (third generation) wireless networks, HiperLAN/2 may have an edge throughout Europe and possibly Asia. HiperLAN is only of interest if you are a frequent international traveler, in which case, you may want to consider buying a HiperLAN wireless adapter for use while abroad.

◆ ◆ ◆

Glossary

802.11 The base standard in the family of IEEE WLAN standards. 802.11 includes three primary physical layer standards; 802.11a, 802.11b, and 802.11g.

802.11a 802.11a is one of the three primary Wi-Fi standards. 802.11a devices have a maximum capacity of 54 Mbps and operate in the unregulated 5 GHz radio band. 802.11a devices do not interoperate with 802.11b or 802.11g devices because they operate on a different frequency.

802.11b Sometimes called *Wireless B,* 802.11b is one of the three primary Wi-Fi standards. 802.11b devices have a maximum capacity of 11 Mbps and operate in the unregulated 2.4 GHz radio band. Because there is an abundance of inexpensive 802.11b equipment, it is still a popular standard. However, it is losing ground to faster 802.11g devices.

802.11g Sometimes called *Wireless G,* 802.11g is one of the three primary Wi-Fi standards. 802.11g devices have a maximum connection speed of 54 Mbps, although in practice they rarely achieve more than 30 Mbps. 802.11g devices operate in the unregulated 2.4 GHz radio band and are backward compatible with 802.11b devices.

802.11i A new security supplement to 802.11, 802.11i addresses security holes in the Wi-Fi standards, and improves encryption (replaces WEP), key management, distribution, and user authentication. The Wi-Fi Alliance based the new Wi-Fi encryption standard, *Wi-Fi Protected Access* (WPA), on an early draft of 802.11i.

802.11x 802.11x is a generic reference used by authors and journalists when referring to all of the Wi-Fi standards. In this book, I use it interchangeably with Wi-Fi to refer to 802.11b, 802.11a, and 802.11g.

access point Access points are Wi-Fi transceivers that act as hubs or routers and connect clients in a wireless network operating in *infrastructure mode.* Access points can also provide a point of connection between a wireless network and a wired LAN.

ad hoc mode Devices in an ad hoc network communicate directly between one another and not through an access point. Ad hoc networks are slower and less reliable than infrastructure mode networks (using access points). Ad hoc is only suited for small groups of clients or to provide temporary connectivity between notebook computers.

antenna A component in a transceiver that directs incoming and outgoing radio waves. Antennas are inactive components and do not create or add power to a broadcast. An antenna can increase signal gain only by concentrating and directing a signal, much like a lens concentrates light to produce a brighter beam.

Asymmetric DSL (ADSL) ADSL is the most common type of DSL and transmits data on analog phones lines. Because the download speed is greater than the upload speed, ADSL is asymmetric.

attenuation A loss of signal power. Attenuation can occur naturally as the distance between transmitter and receiver increases. Interference from obstacles in the signal path, such as building materials, trees, and even people, can increase attenuation, especially in low-power signals.

Basic Service Set (BSS) An access point and one or more associated wireless devices communicating in *infrastructure mode*.

Bluetooth A *Wireless Personal Area Networking* (WPAN) technology developed that operates in the 2.4 GHz radio band. Developers named Bluetooth for the Danish King Harald Blåtand, who unified Denmark and Norway in the tenth century. Bluetooth is a complementary standard to 802.11x, replacing peripheral cables while Wi-Fi replaces Ethernet. Bluetooth's range is too short and its throughput too slow to compete with Wi-Fi.

cantenna A homemade Wi-Fi antenna design that uses a recycled can. The Pringles cantenna is one popular version. Cantennas are surprisingly effective and easy to make.

Denial of Service (DoS) An attack that overwhelms a computer or service and prevents legitimate users from accessing it.

Dynamic Host Configuration Protocol (DHCP) DHCP is a protocol used to assign IP addresses to network clients automatically, each time they connect to the network; for example, after a reboot, or when reconnecting a laptop to the network.

Ethernet The IEEE 802.3 standard for wired networks that supports transmission speeds of up to 100 Mbps.

firewall A hardware device or software application that acts as a traffic filter between internal and external networks. Firewalls prevent unauthorized traffic from entering a network, block access by unauthorized users, and are able to prevent a number of different attacks.

Fresnel zone Pronounced *fray-nel* and named after French physicist Augustin-Jean Fresnel. The Fresnel zone refers to the pattern of radio waves between a transmitter and a receiver. The Fresnel zone between two antennas is shaped like an elongated ellipse, with the widest distribution of RF radiation occurring halfway between transceivers. If the Fresnel zone is not free of obstructions, including trees, the signal power will weaken (*attenuate*) significantly.

Global Positioning System (GPS) A network of 24 satellites used for precise navigation. The GPS satellites circle the earth and transmit time-coded signals. In order to triangulate a position on Earth, a GPS receiver must receive a signal from at least 3 of the 24 satellites. If a device receives a signal from at least 4 satellites, it can determine the user's altitude in addition to their latitudinal and longitudinal position.

HomeRF A wireless home networking standard, now defunct and unsupported, that also operated in the 2.4 GHz band.

infrastructure mode One of the two modes of WLAN operation. In infrastructure mode, WLAN clients communicate through access points, rather than directly with one another.

Institute of Electrical and Electronics Engineers (IEEE) An international, technical professional organization (also called I-triple-E). The IEEE is a nonprofit organization and produced the 802.11 family of standards, as well as standards for Ethernet. The IEEE is one of the leading authorities in many areas, including computers, telecommunications, and power generation.

Internet Connection Sharing (ICS) Windows software that allows you to share your computer's Internet connection with computers on the same network. To share its Internet connection via ICS, a computer needs two network connections, one to the Internet and one to the local network.

Internet Protocol (IP) address A numeric address that every computer must have in order to communicate on a TCP/IP network. The IP address can be dynamically assigned by a DHCP server, or be manually assigned and permanent (static).

Internet Service Provider (ISP) A company that provides Internet access to its customers, either through dial-up connections over the phone lines or through broadband connections (cable and DSL).

Local Area Network (LAN) A network comprised of connected computers in the same local area, for example, an office building or your home.

Line-of-Sight (LOS) A wireless connection that requires an unobstructed path between sending and receiving antennas on a network. The signal must travel from point to point without encountering any obstacles.

MAC address The unique identifier assigned to every network device when it is manufactured. The MAC is usually *hard-coded* and cannot be changed.

Media Access Control (MAC) layer The protocol that defines how network adapters access the physical transmission medium of the network in the case of WLAN radio frequencies.

multipath signal interference Multipath refers to a radio signal that because of obstructions or reflecting off of objects in the environment, splits into multiple signals that arrive at their destination at different times. This can cause problems as these different signals can cancel one another out and cause data loss.

Network Address Translation (NAT) NAT translates internal network IP addresses into one or more public Internet IP addresses, allowing several computers on your network to share a common public IP address. NAT is how a broadband router shares your internet connection with the rest of the network. The router has a public Internet address and acts as a *proxy* between your network clients and the Internet. This protects your internal network and conserves the limited supply of public IP addresses.

Network Interface Card (NIC) A network adapter, usually a PCI card, that allows a computer to connect to the local network, or Wi-Fi network in the case of a wireless NIC.

network latency The amount of time it takes for data to make a round trip from one point to another on a network or across the Internet. Engineers often refer to latency as the *round trip time* (RTT). Many factors affect latency including network traffic levels, distance, signal loss, and interference.

network throughput The amount of information that can be transferred between nodes on a network in a specific amount of time. Advertisers often advertise the potential throughput as a device's speed. For example, 802.11g devices have an advertised maximum throughput of 54 Mbps, when in practice the actual network throughput seldom exceeds 30 Mbps.

Non-Line-of-Sight (NLOS) A NLOS connection doesn't require a *line-of-site,* obstruction-free signal path. NLOS signals can take different routes to their destination, including bouncing or reflecting off of environmental objects.

omnidirectional antenna An antenna that radiates RF signals in all directions, rather than focusing them in one direction. This is the most common type of antenna and is usually the type that comes stock on most access points.

packet switching The way that data is routed on a TCP/IP network (like the Internet). Computers divide data into smaller, individually addressed packets. Because each of these data packets contains a destination address, they can follow different paths to reach their destination. A packet-switched network uses network bandwidth more efficiently because users can send packets at the same time and share bandwidth. Routers direct each individually addressed packet to its proper destination.

Personal Digital Assistant (PDA) A small handheld computer that stores personal information, and has productivity-enhancing applications. The best PDAs are as powerful as the desktop computers in use five to eight years ago. With the addition of a wireless adapter, most PDAs can connect to Wi-Fi or cellular networks.

piconet An ad hoc network created when Bluetooth devices connect. A piconet can have up to seven Bluetooth devices, and piconets can connect to form larger *scatternets*.

Power over Ethernet (PoE) A method of delivering power to Ethernet devices using Ethernet cable rather than an electrical cord. It's possible to deliver power as well as data over Ethernet cable because it has two pairs of wires (4/5 and 7/8) that aren't used in data transmission. PoE equipment uses these wires to supply power to PoE-compatible Ethernet equipment. Using PoE, you can install Ethernet equipment where it's needed, rather than next to power outlets.

Radio Frequency (RF) Electromagnetic frequencies higher than audio frequencies but lower than the frequencies of visible light. Wi-Fi devices use radio frequencies to communicate with each other.

radio frequency channel A section, or slice, of a radio frequency band assigned for communication between two or more devices. Wi-Fi devices have several channels available for use in the 2.4 GHz range.

reversed polarity connectors Also known as *reversed gender* connectors, RP connectors are coaxial plugs and jacks that have the center pin reversed, when compared to standard connectors of the same type.

RF interference Radio frequency noise or competing transmissions in the same frequency band as a WLAN. RF interference can prevent normal operation of a wireless network.

Roundtrip Time (RTT) A term used interchangeably with *latency* to describe the amount of time it takes for data to make a round trip between nodes on a network.

router A device that forwards data traffic between networks. A broadband router forwards data between the Internet and your network clients (see *network address translation*). Some routers also include an integrated *firewall* to protect your network from unauthorized access and attack.

satellite broadband Internet service provided via a satellite modem. Signals travel from the user to the *ISP* (and return) through a satellite in orbit. Early satellite Internet services systems used a phone line for sending data, reserving the satellite connection for downloading web pages and data.

Service Set Identifier (SSID) A name used to identify a WLAN, and distinguish it from other WLANs operating in the same area. An SSID can be up to 32 characters in length.

Software Access Point (SAP) A computer running software that allows it to act as an access point in an *infrastructure mode* wireless network. A software access point duplicates the functionality of a stand-alone hardware access point.

SOHO (small office/home office) An acronym used to identify the consumer market for products that are intended individuals working from home offices or for small businesses.

spoofing When a device, program, or individual uses a false address or authentication information to gain entry to a server or network client. For example, to gain access to a WLAN, a cracker can configure his wireless adapter to spoof the MAC address of an authorized client's adapter. Using that client's Mac address the cracker can pretend to be that user, and access network computers and services.

TCP/IP The principal suite of networking protocols used for communicating on the Internet. Although, TCP/IP is a suite composed of many protocols, its name reflects the two that are the most important: the Transmission Control Protocol (TCP) and the Internet Protocol (IP).

transceiver A term created by combing the words *transmitter* and *receiver*. A transceiver is a two-way radio, a device that transmits and receives radio frequency waves. A WLAN is comprised of transceivers, including access points and client adapters.

Virtual Private Network (VPN) A method of using encryption to protect data and to allow organizations to use the public Internet for communication as if it were a private LAN. In effect, a VPN creates an encrypted tunnel between two points on the Internet. Even though it's sent across a public network, data sent through a VPN connection stays private because of the encryption and authentication.

warchalking Marking a wall or sidewalk with symbols that indicate the presence of an open Wi-Fi access point and provide connection instructions. The idea of warchalking was inspired by the graffiti that American hobos used during the Great Depression to leave messages for one another.

wardriving The act of driving around a geographic area to locate Wi-Fi networks and unsecured access points. Wardrivers use wireless notebook computers or PDAs with high-gain antennas and sniffing software to find WLANs. They often use GPS gear and software to map networks that they find.

Wi-Fi (wireless fidelity) The consumer-friendly name used when referring to any of the three primary 802.11 physical layer protocols (802.11a, 802.11b, and 802.11g). The Wi-Fi alliance (formerly the 802.11 alliance) coined the term Wi-Fi to promote and market 802.11 wireless products more effectively.

Wi-Fi Alliance Founded in 1999, the Wi-Fi Alliance is a membership organization that promotes the use of 802.11 compatible gear and certifies product compliance to the 802.11x wireless standards. If a network component carries the Wi-Fi-certified logo, it will interoperate with other Wi-Fi-certified products that use the same frequency.

Wi-Fi Protected Access (WPA) A new security protocol based on an early draft of the 802.11i protocol. WPA addresses weaknesses in the WEP protocol and gives WLANs better protection.

Wired Equivalent Privacy (WEP) The original security protocol designed to give WLANs protection and privacy equivalent to Ethernet. WEP failed due to design and implementation flaws.

wireless adapter A network interface card (NIC) or adapter that enables a computer or PDA to communicate on a WLAN.

wireless bridge An access point that can communicate with other access points as well as network clients. True wireless-to-wireless bridging is only possible because of the *Wireless Distribution System* (WDS) standard.

Wireless Distribution System (WDS) The standard that allows traffic to flow from one access point to another as if it were traveling from one Ethernet port to another Ethernet port on a wired network. WDS administers to bridge wireless networks as if they were wired networks.

Wireless Local Area Network (WLAN) Rather than using Ethernet cables as a wired LAN does, a WLAN uses radio signals, in the 2.4 GHz and 5 GHz band for communication.

Wireless Metropolitan Area Network (WMAN) A wireless network that encompasses a large geographic area, like a campus, city, or town.

Wireless Personal Area Network (WPAN) A wireless network that serves a single individual or small group. WPANs are short-ranged, and low power; a *piconet* is a form of WPAN. Another example is a computer and printer that are connected with Bluetooth adapters instead of cables, or a group of Bluetooth-enabled PDAs connected in a piconet.

wireless repeater An access point that repeats the signal from another access point, extending the signal range of a WLAN. A repeater doesn't communicate with clients, it merely extends the range of another access point. You can configure many access points to act as repeaters.

Wireless Service Provider (WSP) A company that provides wireless communication services, including cellular, Wi-Fi, Internet, and satellite.

wireless sniffer An application that captures and examines wireless network traffic. You can use sniffers to locate and identify WLANs. Many wardrivers use sniffers.

Wireless User Group (WUG) An organization of wireless technology users that get together and share information to improve their knowledge of wireless technology and share resources.

World Wide Web A service available over the Internet that delivers Web pages to your Web browser. The Internet and the World Wide Web are two different things: The Internet is the network that moves information; the Web delivers the content via your Web browser.

Index

802.11, 16, 249
802.11a, 17–18, 244, 249
802.11b, 16, 243, 249
802.11c, 244
802.11d, 244
802.11e, 245
802.11f, 245
802.11g, 17, 244, 249
802.11h, 245
802.11i, 17, 245, 249
802.11n, 18, 246
802.11 IR, 246
802.15.1, 18–20, 247
802.15.3, 248
802.15.4, 247
802.16a, 246
2600 Magazine, meetings, 54
2600: The Hacker Quarterly, 58

A

access control, passwords and, 169
access points
    accidental connections and, 129
    bridge mode, 11–12
    client mode, 11–12
    configuration errors, 132–133
    definition, 249
    description, 9
    DHCP and, 11, 193
    hubs, 11
    interoperability and, 138
    IP addresses and, 191–192
    network settings, 193
    normal mode, 11–12
    print server, 11
    repeater mode, 11, 13
    rogue access points, 68–70
    routers, 11
    switches, 11
    WPA, protection and, 182
accessory manufacturers, 240–241
accidental connections, 129
active devices, hubs, 5
ad hoc mode, 10, 250
ad-blocking software, 153

adapters
    built-in adapters, 13–14
    CF (Compact Flash) card adapters, 13–14
    modes, configuration errors and, 133
    PCI adapter cards, 13–14
    PCMCIA card adapters, 13–14
    USB, 13–14
address books, worms and, 113
addresses
    classes, 35–36
    DHCP and IP, 38–39
    IP security, 191–193
    restricted, 36
    subnet masks, 38
    subnets, 36–37
adjacent channels, overlap, 130
ADSL (Asymmetric DSL), 250
adware. *See* spyware
antennas
    decibels, 140
    definition, 250
    Fresnel zone, 143–144
    gain, 140
    grounded, 142–143, 229–230
    installation, configuration errors and, 133
    interference and, 127
    introduction, 138
    mounts, 140–142
    RF signals, 139
    transceivers and, 139
    trees and, 144–145
antispyware software, 116, 152
antivirus software
    adware/spyware and, 153
    drive-by downloads and, 152
    hotspot protection and, 65
    resource Web sites, 238–239
    updates, 213
    worms, 114
Application layer
    OSI reference model, 27–28
    TCP/IP model, 29, 30
ARP (Address Resolution Protocol), 32
attenuation
    definition, 8, 250
    Fresnel zone and, 144
    inline attenuators, broadcast power and, 203
    sniffers and, 85

B

backdoors, Trojans and, 111
backups
    bare metal, 220
    differential, 220
    frequency, 220
    incremental, 220
    Internet, 225–226
    media selection, 223–224
    nearline storage, 222–223
    offline storage, 223
    online storage, 220–221
    planning considerations, 224–225
    software, 226–228
    storage, 220–223
    what to back up, 219
bandwidth, sharing, 131
bare metal backups, 220
BAT file extensions, viruses and, 107
beacon frames, wardriving and, 91
black hats, 46–47
Blaster worm, 48
blended attacks, spoofing and, 65
blended threats, 101, 114–115
Bluetooth
    802.15.1 and, 18–20
    definition, 250
    master devices, 18
    piconets, 18
    scatternets, 19
    slave devices, 18
Bots, 101, 116–117
bridge mode, access points, 11–12, 11–13
bridges, network connection and, 41
British government, hacktivist, 49
broadcast
    connection speed, minimum, 205
    inline attenuators, 203
    overview, 10
    parameter modification, security, 202–205
    power adjustments, 203
    SSID broadcast, disabling, 204–205
browsing, VPN and, 67
BSS (Basic Service Set), 250
built-in adapters, 13

C

cabinets, interference and, 127
cameras, 159, 161–163
cantenna, 250

case sensitivity, SSID, 190
ceilings, interference and, 127
CERT (Computer Emergency Response Team), security vulnerabilities, 85
certificate authority, digital certificates and, 177
CF (Compact Flash), card adapters, 13–14
channels
    adjacent, overlap, 130
    configuration errors and, 133
Chaos Computer Club, 58
ciphers
    definition, 168
    encryption key, 170
    frequency analysis, 172
ciphertext, 170
classes, addresses, 35–36
client mode, access points, 11–12
clients
    description, 26
    excess number, 133
    introduction, 4
    security, 207–215
    wireless adapters and, 9
    wireless NICs and, 9
coaxial cable, Ethernet networks and, 5
CodeRed worm, 115
COM file extensions, viruses and, 107
Computer Browser service, 214
configuration
    adapter modes, 133
    antenna installation and, 133
    channels, 133
    connection speed, 133
    error checking, 132–133
    IP addresses, 133
    SSID and, 133
connections
    accidental, 129
    bridges, 41
    gateways, 41–42
    routers, 41–42
    speed, configuration errors and, 133
    speed, minimum, 205
    unauthorized, identifying, 131
conventions/gatherings for hackers and crackers, 53–54
cookies, privacy and, 154–156
cordless phones, 163–165
CPUs (central processing units), hubs and, 6

crackers
    2600 Magazine, 54
    black hats, 47
    conventions/gatherings, 53–54
    digital certificates and, 178
    fink-fund virus reporting reward, 109
    hacker comparison, 45–46
    jargon, 50–51
    packet monkeys, 47
    reasons for cracking, 51–53
    resources, 57–58
    script kiddies, 47–49
    session hijacking, 59
    s'kiddiots, 47
    social engineering, 50, 55–56
    spoofing and, 60
    trashing, 56
    tricks, 54–57
    viruses and, 102
cross-contamination of infection, 103
cryptanalysis, 168
cryptography, 168, 173–176
cryptology, 168
cryptosystems, 168
cyber vandalism, script kiddies and, 48

D

data files, viruses and, 107
Data link layer, OSI reference model, 29
Data link (MAC) layer, TCP/IP model, 29, 30
data protection, backup systems, 219–228
data rate, 8
data threats, 217–219
DDoS (distributed denial of service) attacks, 70
decibels, antenna gain, 140
decryption, 168
DEFCON hacking convention, 53
demultiplexing, encapsulation and, 34
DHCP (Dynamic Host Configuration Protocol)
    access points and, 193
    definition, 250
    disabling, 193
    ICS (Internet Connection Sharing) and, 133, 194
    introduction, 32
    IP addresses and, 38–39
    IP addresses assignment, 193
    problems, 194
    security and, 193–197
    services, access points and, 11
    wardriving and, 94–95
differential backups, 220

digital certificates, 177–178
digital envelope, public key systems, 175
digital signatures, one-way-hash function, 176–177
directional antennas, 133
disassociate frames, 61, 73
disassociation messages, DoS attacks, 61
distributed denial of service attack (DDoS), 70
diversity antenna system, multipath interference and, 127
DMZ (demilitarized zone), security and, 210–211
DNS (Domain Name System), 32
DoC (Dis Org Crew), 58
domain names, 39–40
DoS (denial of service) attacks
    accidental, 71
    bots and, 116
    definition, 250
    disassociate frames, 73
    disassociation messages and, 61
    exposing, 70–74
    FakeAP, 74
    networking device flaws and, 71
    session hijacking and, 61
    strong signal jamming, 73–74
    WPA, 72–73, 182–183
drive-by downloads, 152–154
drivers, optimization and, 146
DSS (Digital Spread Spectrum), cordless phones and, 164
DSSS (Direct Sequence Spread Spectrum), interference and, 129
dual-band devices, 16
dust accumulation, hardware failure and, 218

E

ego, hackers/crackers and, 52
electromagnetic waves, antennas and, 139
e-mail, social engineering and, 56
EMF (electromagnetic field), detection, 136
employees, crackers and, 55
encapsulation, 33–34
encryption
    cipher, 168
    cipher frequency analysis, 172
    ciphertext, 170
    cordless phones, 164
    cryptography, 168
    decryption, 168
    history of, 171–172
    hotspot protection and, 65

Continued

encryption *(continued)*
    hybrid systems of cryptography, 175–176
    introduction, 167
    modern techniques, 173–179
    NetStumbler, 96
    passwords and, 169
    patterns, 172
    PGP (Pretty Good Privacy), 184
    plaintext, 168–169
    public key cryptography, 174–175
    scytale, 171
    security and, 10, 205–206
    steganography, 171
    transposition cipher, 171
    VPNs and, 173
    wardriving protection and, 95–96
    WEP and, 96, 179–181, 205–206
    WPA and, 96, 182
encryption key
    ciphertext, 170
    description, 170
    entropy and, 178–179
    history, 172–173
    key length, 178–179
    pseudo-random key, 180
    WEP and, 181
entropy, encryption key length and, 178–179
equipment
    access points, 11
    adapters, 13–14
    to avoid, 14
    cost comparison, 7
    optimization, 145–146
    standards, 14–15
    upgradeable, 15
errors in configuration, identification, 132–133
ESD (electrostatic discharge), hardware failure and, 218
Ethernet networks
    coaxial cabling and, 5
    definition, 250
    equipment cost, 7
    hybrid with wireless, 5
    security, 8
    speed compared to Wi-Fi, 6, 8
    wireless comparison, 5–8
EXE file extensions, viruses and, 106

F

FakeAP DoS attack, 74
FCC, hoaxes and, 121
file extensions, viruses and, 106–108

file-infecting macro viruses, example, 107
filtering
    IP addresses, implementation, 201–202
    MAC addresses, activation, 199–201
    security and, 197–202
    wardriving protection and, 95
fink-fund, virus reporting reward, 109
firewalls
    definition, 250
    drive-by downloads and, 152
    hotspot protection and, 65
    ICF (Internet Connection Firewall), 208–210
    personal, 79
    spoofing and, 64
    spyware and, 119
    XP setup, 208–210
firmware
    optimization and, 146
    upgradeable, 15
    WEP and, 181
    WPA, upgrades, 206
flash chips, optimization and, 146
floors, interference and, 127
frequency analysis, ciphers, 172
Fresnel zone
    antennas and, 143–144
    definition, 251
FTP (File Transfer Protocol)
    description, 32
    ports, 78
furniture, interference and, 127

G

gain, antennas, 140
gateways, connections and, 41–42
Google, privacy and, 148
GPS (Global Positioning System)
    definition, 251
    wardriving and, 98
gray hats, security and, 47
grounded antennas, 142–143, 229–230

H

hackers
    2600 Magazine, 54
    conventions/gatherings, 53–54
    cracker comparison, 45–46
    MIT and, 45–46
    reasons for hacking, 51–53
    resources, 57–58
hacktivists, 49–50

handhelds, antivirus software, 120
hardware. *See also* firmware
    failure, 218
    grounding antennas, 229–230
    lightening arrestors, 229–230
    maintenance, protection and, 230–232
    manufacturers, 233–236
    protection tips, 228–232
    SSID and, 190
    static electricity, 231
    surge protection, 228–232
hijacking, session hijacking, 59
HiperLAN (High-Performance Radio Local Area Network), 248
hoaxes
    description, 120–121
    FCC and, 121
    identification, 121–122
    Trojans and, 121
hobo symbols, 97
home networks
    attacks, 77
    TCP/IP port scanners, 78
HomeRF working group, 14–15, 251
HOPE (Hackers on Planet Earth) convention, 53
host names, 40
hosts
    definition, 26
    security, 207–215
hotspots
    protection in, 65–67
    rogue access points and, 68
    session hijacking, 64–65
HTA file extensions, viruses and, 107
HTTP (Hypertext Transfer Protocol)
    description, 32
    proxy servers, VPNs and, 67
hubs
    access points and, 11
    as active devices, 5
    CPUs and, 6
    multiport repeaters, 6
    as passive devices, 5
    routing and, 6
hybrid cryptosystems
    digital certificates, 177–178
    digital signatures, 176–177
    overview, 175–176
hybrid wireless/Ethernet network, diagram, 5

I

ICF (Internet Connection Firewall) in XP, setup, 208–210
ICMP (Internet Control Message Protocol), 32
ICS (Internet Connection Sharing)
    access point configuration and, 133
    definition, 251
    DHCP disabling and, 194
IEEE (Institute Electrical and Electronics Engieers), 20–21, 251
IIS (Internet Information Server), 214
incremental backups, 220
infections. *See* viruses
infrastructure mode, 10, 251
interference
    cabinets and, 127
    ceilings and, 127
    cordless phones, 134, 135
    detection, 134–137
    devices that cause interference, 134
    DSSS and, 129
    floors and, 127
    furniture and, 127
    home sources, 135
    identification, 135–136
    microwave ovens, 134
    multipath, 125–130
    neighbors' setup, 135–136
    OFDM and, 129
    office sources, 135
    outside sources, 135–136
    physical barriers, 136–137
    radio waves and, 126
    RF, detection, 136
    site surveys and, 129
    trees, 144–145
    walkie-talkies, 134
    walls and, 127
    wireless cameras, 134
    wireless speakers, 134
Internet
    backups, 225–226
    definition, 25
Internet Explorer, privacy settings, 154–156
Internet Systems Consortium, 111
Internet transponders, drive-by downloads and, 152–153
internetwork, 25
interoperability, 138
intranet, 25

IP addresses
    access point information and, 191–192
    conflicts, configuration errors and, 133
    default settings, wardriving protection and, 94
    definition, 251
    description, 35
    DHCP and, 38–39
    DHCP assignment of, 193
    domain names and, 39–40
    filtering, implementation, 201–202
    host names, 40
    manual, 133
    root domains and, 40
    security and, 191–193
    static, security and, 195
    subdomains, 40
    VPNs and, 66
IRC (Internet Relay Chat), script kiddies and, 49
ISPs (Internet Service Providers), 251
IV (initialization vector), WEP and, 180

J
jargon of crackers, 50–51
JS file extensions, viruses and, 107
jukeboxes, nearline storage of backups, 223

K
keyloggers, 119

L
LANs (local area networks)
    definition, 251
    flexibility of Wi-Fi and, 4
    rogue access points and, 68
latency, 8
layered defense, blended threats, 115
legal issues, wardriving, 90, 96–97
license agreements, spyware and, 118
life-cycle of virus, 105
lightning arrestors, hardware protection and, 229–230
LNK file extensions, viruses and, 107
location-based advertising, 149–151
LOS (Line-of-Sight), 251

M
MAC (Media Access Control) address
    definition, 251
    filter activation, 199–201
    filters and, 197–198
    wardriving protection and, 95
MAC (Media Access Control) layer, Wi-Fi, 252

macro viruses
    executable programs, 104
    file-infecting, 107
malware, 101–102
man-in-the-middle attacks. *See* MITM (man-in-the-middle) attacks
manufacturers, hardware, 233–236
masks, subnets, 38
master devices, Bluetooth and, 18
Media access control layer, Wi-Fi, description, 31
media for backups, 223–224
Microsoft, fink-fund virus reporting reward, 109
microwave ovens, interference and, 134
MIT (Massachusetts Institute of Technology), hackers and, 45–46
MITM (man-in-the-middle) attacks
    exposing, 68
    session hijacking and, 61
Morris Internet Worm, 111
mount antennas, 140–142
multipath interference
    antenna mounts, 142
    antennas, 127
    cabinets, 127
    ceilings, 127
    definition, 252
    description, 125
    DSSS and, 129
    floors, 127
    metal furniture, 127
    OFDM and, 129
    radio waves and, 126
    site surveys and, 129
    walls, 127
    WLANs and, 127
multiport repeaters, active hubs as, 6

N
NAT (network address translation), 42, 252
nearline storage of backups, 222–223
neighbors' setup, interference and, 135–136
Netmeeting Remote Desktop Sharing, 214
NetStumbler, encryption and, 96
Network infrastructure layer, Wi-Fi, 31
network latency, 252
Network layer
    OSI reference model, 29
    TCP/IP model, 29, 30
network throughput, 252
networks. *See also* home networks
    clients, 26
    connections, 40–43
    default settings, wardriving and, 92–94

definition, 24
device flaws, DoS attacks and, 71
host, 26
host attacks, 77–85
Internet, 25
internetwork, 25
intranet, 25
models, 27–32
nodes, 24
OSI reference model, 27–29
peers, 27
protocols, 24
servers, 26
services, 26
speed, overview, 8
subnets, 36–37
terminology, 23–27
topologies, 24
traffic filtering, 197–202
wardriving and, 91–92
workstations, 27
NIC (Network Interface Card)
clients, 9
definition, 252
promiscuous mode, 87–88
sniffers and, 87–88
Nimda worm, 115
NLOS (Non-Line-of-Sight), 252
nodes, 24
normal mode, access points, 11–12

O

OFDM (Orthogonal Frequency Division Multiplexing), interference and, 129
offline storage of backups, 223
omnidirectional antenna, 252
one-way-hash function, digital signatures, 176–177
online storage of backups, 220–221
open protocols, 24
organizations, Wi-Fi Alliance, 20–21
OS (operating system)
lockdown, 208–213
updates, adware/spyware and, 153
weaknesses in, 79–82
OSI (Open Systems Interconnect) reference model, 27–30

P

packet monkeys, 47
packet switching, 33–34, 252
packets, 6, 61
Parson, Jeffrey Lee, 48

passive devices, hubs, 5
passwords
access control and, 169
encryption and, 169
selection tips, 190–191
social engineering and, 55–56
SSID and, 190
trashing, crackers and, 56
wardriving protection and, 93
WEP and, 181
patches, security and, 211–213
payloads, viruses, 104, 111
PC manufacturers, 240–241
PCI, adapter cards, 13–14
PCMCIA, card adapters, 13–14
PDAs (personal digital assistants)
antivirus software, 120
as clients, 4
definition, 253
manufacturers, 240–241
virus infection, 103–104
peers, networks, 27
personal firewalls, 79. See also firewalls
PGP (Pretty Good Privacy), public key cryptography, 175, 184
physical barriers, interference and, 136–137
Physical layer
OSI reference model, 29
Wi-Fi, 31
piconets, 18, 253
PIF file extensions, viruses and, 107
PLA (Phone Losers of America), 58
plaintext
ciphertext, 170
in the clear, 169
encryption and, 168–169
PoE (Power over Ethernet), 253
policies on privacy, 156–157
port scanners
home network attacks, 78
services check, 79
unauthorized connections and, 131
power surges, hardware failure and, 218
P3P (Platform for Privacy Preferences), 157
Presentation layer, OSI reference model, 29
print servers, access points and, 11
printers, as clients, 4
privacy. See also encryption
cookies and, 154–156
cordless phones, 163–165
Google and, 148
PGP (Pretty Good Privacy), 184

Continued

privacy *(continued)*
  policies, 156–157
  P3P, 157
  public databases and, 148
  reasons for, 147–149
  resource Web sites, 238–239
  spyware exposure, 151–152
  threats, 149–156
  video cameras, 161–163
  WEP, 179–181
  WPA and, 182–183
privacy settings, Internet Explorer, 154–156
promiscuous mode, NICs, 87–88
proprietary protocols, 24
protection tips
  backups, 219–228
  blended threats, 115
  bot software, 117
  data threat identification, 217–219
  hardware protection, 228–232
  public hotspots, 65–67
  Trojans, 111
  worms, 114
protocols
  definition, 24
  open, 24
  proprietary, 24
  TCP/IP, 32
proxy servers, HTTP, VPNs, 67
pseudo-random encryption key, 180
public databases, privacy and, 148
public hotspots, session hijacking, 64–65
public key cryptography
  digital envelope, 175
  invention, 174
  PGP (Pretty Good Privacy), 175

R

race conditions
  overview, 64
  spoofing and, 61
radio bands, broadcast bands, 4
radio frequency channel, 253
radio waves, interference and, 126
RAID, online backup storage, 221–222
Remote Desktop Help Session Manager, 215
Remote Registry, 215
repeater mode, access points, 11, 13
restricted addresses, 36
revenge, hackers/crackers and, 52–53
reversed polarity connectors, 253

RF interference
  definition, 253
  detection, 136
RF (Radio Frequency), 253
RF signals, antennas and, 139
rogue access points
  hotspots and, 68
  LANs and, 68
  WEP encryption and, 70
  WPA encryption and, 70
root domains, IP addresses and, 40
ROT13 encryption, 172
routers
  access points and, 11
  connections and, 41–42
  definition, 253
  hubs and, 6
  nodes, 24
  spoofing and, 64
routing, hubs, 6
Routing & Remote Access, 215
RTT (Roundtrip Time), 253

S

SAP (Software Access Point), 254
satellite broadband, 253
scan lines, television, 126
scatternets, Bluetooth and, 19
SCR file extensions, viruses and, 107
script kiddies, 47–49
scytale, 171
secret key cryptography, failings, 173–174
security
  antivirus software, 213
  backup systems, 219–228
  black hats, 46, 47
  Blaster worm, 48
  broadcast and, 10
  broadcast parameters modification, 202–205
  CERT Web site, vulnerabilities, 85
  clients, 207–215
  cyber vandalism, 48
  data threat identification, 217–219
  default settings and, 188–193
  DHCP and, 193–197
  DMZ (demilitarized zone), 210–211
  encryption and, 10, 205–206
  Ethernet networks, 8
  firewall setup in XP, 208–210
  gray hats, 47

hacktivists, 49–50
hosts, 207–215
IP addresses, 191–193
IP addresses, static, 195
known issues exploitation, 83–85
network traffic filtering, 197–202
password selection tips, 190–191
patches, 211–213
resource Web sites, 238–239
script kiddies, 47–49
services, disable unnecessary, 214–215
updates, 211–213
urban locations, 188
white hats, 46
WLAN vulnerabilities, 187–188
WPA firmware upgrades, 206
Server Service, 215
servers, overview, 4, 26
services
    disable unnecessary, 214–215
    networks, 26
    theft, identification, 131–132
session hijacking
    definition, 59
    diagram flow charts, 62
    DoS attacks, 61
    MITM (man-in-the-middle) attacks, 61
    public hotspots, 64–65
    time outs and, 59
    user sessions, 59
Session layer, OSI reference model, 29
SHS file extensions, viruses and, 107
signal jamming, DoS attack, 73–74
signal propagation, multipath, 126
signals, X10 device signal interception, 158–161
Simple File Sharing, vulnerabilities and, 81
site surveys, multipath interference and, 129
s'kiddiots, 47
slave devices, Bluetooth and, 18
SMTP (Simple Mail Transfer Protocol), ports, 78
sniffers
    attenuation and, 85
    NICs and, 87–88
    software, 86–87
    spoofing and, 60
social engineering, crackers and, 50, 55–56
software. *See also* firmware
    ad-blocking, 153
    backup software, 226–228
    bot software, 116–117
    malicious, 101

spy-blocking, 153
wardriving defense, 99
Web sites, 241–242
SOHO (small office/home office), 254
speed
    attenuation, 8
    data rate, 8
    latency and, 8
    throughput, 8
    Wi-Fi versus Ethernet, 6
spoofing
    blended attacks and, 65
    definition, 254
    disassociate frames, 61
    firewalls and, 64
    MAC addresses, 61, 63
    race conditions and, 61
    routers and, 64
    sniffers and, 60
spy-blocking software, 153
spycleaners, 119
spyware
    description, 118
    drive-by downloads, 152
    exposure, 151–152
    keyloggers, 119
    system monitors, 119
SSDP Discovery Service, 215
SSID (service set identifier)
    access point location and, 190
    broadcast disable, 204–205
    case sensitivity, 190
    configuration errors and, 133
    creation tips, 190
    default, wardriving protection and, 93, 188–190
    definition, 254
    hardware identification and, 190
    network protection and, 91
    passwords and, 190
    personal information in, 190
    usernames and, 190
standards
    802.11, 249
    802.11a, 244, 249
    802.11b, 243, 249
    802.11c, 244
    802.11d, 244
    802.11e, 245
    802.11f, 245
    802.11g, 244, 249
    802.11h, 245

Continued

standards *(continued)*
    802.11i, 245, 249
    802.11IR, 246
    802.11n, 246
    802.11x, 249
    802.15.1, 247
    802.15.3, 248
    802.15.4, 247
    802.16a, 246
    Bluetooth, 247
    HiperLAN, 248
    introduction, 15
    outdated, 14–15
    WiMAX, 246
    WiMedia, 248
    ZigBee, 247
star network topology, 5
static electricity, hardware and, 231
static IP addresses, security and, 195
steganography, 171
storage, backups, 220–223
strong signal jamming DoS attack, 73–74
subdomains, 40
subnets
    addresses, 36–37
    masking, 38
surge protection, 228–229
switches, access points and, 11
symmetric key cryptography, failings, 173–174
system monitors, 119

T

TCP/IP model
    Application layer, 30
    Data link (MAC) layer, 30
    Network layer, 30
    OSI model comparison, 29–30
    Transport layer, 30
TCP/IP port scanners, home network attacks, 78
TCP/IP protocols
    definition, 254
    descriptions, 32
    encapsulation, 33–34
    packet switching, 33–34
Telnet, 214
theft of service identification, 131–132
threats to privacy, location-based services, 149–151
throughput, 8
time outs, session hijacking and, 59
TLD (top-level domain), 40
ToorCon hacker/cracker convention, 54
topologies, 5, 24

traffic filtering, network security and, 197–202
transceivers
    antennas and, 139
    definition, 254
Transport layer
    OSI reference model, 29
    TCP/IP model, 29, 30
transposition cipher
    encryption and, 171
    ROT13 encryption, 172
trashing, crackers and, 56
trees, interference, 144–145
Trojans
    backdoors and, 111
    description, 110
    hoaxes and, 121
    introduction, 101
troubleshooting
    adjacent channel overlap, 130
    configuration errors, 132–133
    equipment optimization, 145–146
    interference detection, 134–137
    interoperability issues, 138
    multipath interference, 125–130
    theft of service identification, 131–132

U

UDP (User Datagram Protocol), 32
unauthorized connections, identification, 131
Universal Plug and Play Device Host, 214
updates, security and, 211–213
UPS (Uninterruptible Power Supply), power surges and, 229
USB (Universal Serial Bus), wireless adapters, 13–14
Usenet newsgroups, script kiddies and, 49
user group Web sites, 237–238
user sessions, session hijacking and, 59
usernames
    SSID and, 190
    wardriving protection and, 93

V

VBE file extensions, viruses and, 107
VBS file extensions, viruses and, 106
video cameras, 159, 161–163
viruses. *See also* Trojans
    BAT file extensions, 107
    COM file extensions, 107
    cross-contamination, 103
    data files and, 107
    EXE file extensions, 106
    file extensions, double extensions, 107–108

fink-fund virus reporting reward, 109
HTA file extensions, 107
JS file extensions, 107
life-cycle, 105
LNK file extensions, 107
macro, file-infecting, 107
payloads, 104
PDAs, 103–104
PIF file extensions, 107
SCR file extensions, 107
SHS file extensions, 107
sources, 102
VBE file extensions, 107
VBS file extensions, 106
warning hoaxes, 122
worm comparison, 111
VPN (virtual private network)
    definition, 254
    encryption and, 173
    hotspot protection and, 65
    HTTP proxy servers and, 67
    IP addresses and, 66
    Web browsing and, 67

W

walkie-talkies, interference and, 134
walls, interference and, 127
warchalking
    definition, 254
    GPS and, 98
    hobo symbols, 97
wardriving
    beacon frames, 91
    default settings, protection and, 92–94
    defense software, 99
    definition, 254
    DHCP and, 94–95
    encryption and, 95–96
    equipment needed, 91
    filtering and, 95
    GPS and, 98
    IP addresses and, 94
    legal issues, 90, 96–97
    MAC addresses and, 95
    network hiding, 91–92
    passwords and, 93
    SSID broadcast and, 91, 188–190
    usernames and, 93
    viruses and, 102
    wardialers and, 89
    Web sites, 100

    WUGs and, 89
    WWWD (Worldwide Wardrive), 54
warspying, 98–99
wave cycle, electromagnetic waves, 139
wavelengths, electromagnetic waves, 139
WDS (Wireless Distribution System)
    bridging and, 41
    definition, 255
Web sites
    accessory manufacturers, 240–241
    antivirus resources, 238–239
    PC manufacturers, 240–241
    PDA manufacturers, 240–241
    privacy resources, 238–239
    security resources, 238–239
    software, 241–242
    wardriving, 100
    wireless users, 237–238
WEP (Wired Equivalent Privacy)
    802.11i and, 17
    definition, 255
    encryption and, 96, 181
    encryption key and, 181
    failures, 179–180
    firmware and, 181
    IV (initialization vector), 180
    passwords and, 181
    pseudo-random key, 180
    reasons to use, 180–181
    rogue access points and, 70
    security and, 205–206
    video camera encryption, 163
wetware, crackers and, 55
white hats, security and, 46
Wi-Fi
    definition, 254
    equipment costs, 7
    installation, 8
    mobility and, 8
    speed, 6
Wi-Fi Alliance
    definition, 255
    introduction, 3
    overview, 20–21
Wi-Fi layers
    Media access control, 31
    Network infrastructure, 31
    Physical control, 31
WiMAX standard, 246
WiMedia standard, 248

Windows Update, 212–213
Windows XP Home Edition
  clients, static IP address assignment, 195–197
  firewall setup, 208–210
  Guest account password assignment, 82
  ICF (Internet Connection Firewall), 208–210
  ICS (Internet Connection Sharing), 214
  lockdown, 208–213
  NTFS, security and, 214
  passwords and, 214
  physical security, 214
  Simple File Sharing, 81
wireless adapters
  clients and, 9
  definition, 255
wireless bridges, 255
wireless cameras, interference and, 134
wireless networking
  Ethernet network comparison, 5–8
  hybrid network diagram, 5
  overview, 4–5
wireless NICs, clients and, 9
wireless repeaters, 255
wireless sniffers, 255
wireless speakers, interference and, 134
Wireless-G, 802.11a and, 18
WLAN (Wireless Local Area Networking)
  accidental connections and, 129
  ad hoc mode, 10
  attacks, 77
  definition, 4, 255
  infrastructure mode, 10
  multipath interference and, 127
  session hijacking, 59
  signal strength, FCC regulation, 135
  SSID settings, 188–190
  urban locations, 188
  vulnerabilities, 187–188
  worm protection, 115–116

WMAN (Wireless Metropolitan Area Networks), 255
workstations, 4, 27
World Wide Web, 256
worms
  bots and, 116
  CodeRed, 115
  infection cycle, 112
  introduction, 101
  Morris Internet Worm, 111
  Nimda worm blended threat, 115
  protection, 114
  spread methods, 113
  virus comparison, 111
  WLAN security, 115–116
WPA (Wi-Fi Protected Access), 15
  definition, 255
  denial of service, 72–73
  DoS attacks, 182–183
  encryption and, 96, 182
  firmware upgrades, 206
  rogue access points and, 70
  weaknesses in, 206
  WEP and, 182
WPAN (wireless personal area network)
  Bluetooth and, 18
  definition, 255
WPS (Wi-Fi positioning system), 150
WSP (Wireless Service Provider), 255
WUGs (Wireless User Groups), 89, 256
WWWD (Worldwide Wardrive), 54

X

X10 devices, signal interception, 158–161
XP. *See* Windows XP Home Edition

Z

ZigBee standard, 247
zombies, DoS attacks and, 70